Value Management in Construction and Real Estate

T0188669

This cutting edge study explores alternative methods and tools to obtain value for money while maintaining quality in construction projects, especially large and complex ones. Extensive references throughout will help the reader develop a deeper understanding of the methodology, and self-study questions help to keep you on track. This book is ideal as a reference for practitioners and the perfect intro for students of construction or real estate.

Geoffrey Q.P. Shen is an Associate Dean of the Faculty of Construction and Environment and a Chair Professor in the Department of Building and Real Estate of The Hong Kong Polytechnic University. He has 20+ years of experience in the field of value management. He has taught the subject at both undergraduate and postgraduate levels, has undertaken a large number of research projects with total funding over HK$30 million, and has authored 160+ papers in academic journals and 30+ scholarly research monographs. He has professionally facilitated 30+ value management and partnering workshops for many client organisations in both the public and private sectors. He received the Presidential Citation Award from SAVE International in 2009 for his 'energetic and engaging effort to enhance value research and education'.

Ann T.W. Yu is an Associate Professor in the Department of Building and Real Estate of The Hong Kong Polytechnic University. She has 10+ years of experience in the field of value management, teaching in both undergraduate and postgraduate levels, conducting research projects and carrying out consultancy services. She is a Chartered Builder, Quantity Surveyor and Project Manager by profession. She was the Honorary Secretary of the Hong Kong Institute of Value Management for seven years. She has a strong track record and has published extensively on the broad theme of project management in leading construction management journals and international conference proceedings. She received the Highly Commended Award from the Emerald Literati Award for her first paper in Value Management.

Value Management in Construction and Real Estate
Methodology and applications

**Edited by Geoffrey Q.P. Shen
and Ann T.W. Yu**

Sponsored by HKIS

Taylor & Francis Group

LONDON AND NEW YORK

THE HONG KONG INSTITUTE OF
SURVEYORS
香港測量師學會

First published 2016
by Routledge
2 Park Square, Milton Park, Abingdon, Oxon OX14 4RN

and by Routledge
711 Third Avenue, New York, NY 10017

Routledge is an imprint of the Taylor & Francis Group, an informa business

British Library Cataloguing-in-Publication Data
A catalogue record for this book is available from the British Library

Library of Congress Cataloging in Publication Data
Value management in construction and real estate : methodology and
applications / edited by Geoffrey Q.P. Shen and Ann T.W. Yu.
pages cm
Includes bibliographical references and index.
1. Building--Cost effectiveness. 2. Value analysis (Cost control)
I. Shen, Geoffrey. II. Yu, Ann T. W.
TH438.15.V35 2016
690.068'1--dc23
2015017241

ISBN: 978-1-138-85276-1 (hbk)
ISBN: 978-1-138-85278-5 (pbk)
ISBN: 978-1-315-72329-7 (ebk)

Typeset in Times New Roman
by Fish Books Ltd.

Printed and bound in the United States of America by Publishers Graphics,
LLC on sustainably sourced paper.

Contents

Illustrations

Figures

Tables

Contributors

Dr. Jacky Chung is an Assistant Professor with the Department of Building, School of Design and Environment, National University of Singapore. He is an experienced researcher in the areas of Collaborative Team Work, Construction Briefing, Stakeholder Engagement, Public-private Partnership (PPP), Value Management (VM), and Building Information Modeling (BIM). He has been presented six scholarly awards and has produced more than 30 research publications in leading academic journals and international conferences.

Dr. Shichao Fan is the Operation Director of Fosun Property Holdings with the responsibility of re-energizing the acquired projects/buildings in the US. Before joining Fosun Property Holdings, Dr. Fan was a senior analyst of Valuation and Advisory Service of CBRE (HK). Dr. Fan is also a CFA and FRM charter holder. He obtained his Bachelor degree in Civil Engineering from Tsinghua University, and his PhD degree from the Department of Building and Real Estate of the Hong Kong Polytechnic University.

Professor John Kelly was the Morrison Professor of Construction Innovation at Glasgow Caledonian University and is now retired. Having trained and practised as a quantity surveyor before becoming an academic, he had worked in the field of value management for over 20 years, the majority of that time with Professor Steven Male. Together they have published widely, conducted considerable funded research and undertaken value-related consultancy across a wide spectrum of clients and projects. John's major interest lies in the investigation of value strategies at the inception of projects and includes partnering and project briefing.

Dr. Gongbo Lin received his bachelor degree from Tsinghua University in 2003 and his PhD from the Hong Kong Polytechnic University in 2009. He joined Zhejiang Youcheng Group in 2011 where he was the Vice President. Since 2014, he has been with Country Garden Holdings, where he is currently an Assistant Regional Director.

Zuhaili Mohamad Ramly is a Ph.D. candidate at the Department of Building and Real Estate, The Hong Kong Polytechnic University and an academic member of the Department of Quantity Surveying, Faculty of Built Environment, Universiti Teknologi Malaysia (UTM). He holds bachelor degree in quantity surveying and master degree in construction contract, both from UTM.

Geoffrey Q.P. Shen is an Associate Dean of the Faculty of Construction and Environment and a Chair Professor in the Department of Building and Real Estate of The Hong Kong Polytechnic University. He has 20+ years of experience in the field of value management. He has taught the subject at both undergraduate and postgraduate levels, has undertaken a large number of research projects with total funding over HK$30 million, and has authored 160+ papers in academic journals and 30+ scholarly research monographs. He has professionally facilitated 30+ value management and partnering workshops for many client organisations in both the public and private sectors. He received the Presidential Citation Award from SAVE International in 2009 for his 'energetic and engaging effort to enhance value research and education'.

Ann T.W. Yu is an Associate Professor in the Department of Building and Real Estate of The Hong Kong Polytechnic University. She has 10+ years of experience in the field of value management, teaching in both undergraduate and postgraduate levels, conducting research projects and carrying out consultancy services. She is a Chartered Builder, Quantity Surveyor and Project Manager by profession. She was the Honorary Secretary of the Hong Kong Institute of Value Management for seven years. She has a strong track record and has published extensively on the broad theme of project management in leading construction management journals and international conference proceedings. She received the Highly Commended Award from the Emerald Literati Award for her first paper in Value Management.

Preface

This textbook is based on the 20 years of experience in researching, practicing and teaching Value Management (VM). The book provides useful references to help develop a deeper understanding of the methodology among interested readers. It is designated for both undergraduate and postgraduate students in construction and real estate. The book also serves as a good reference for clients and professionals in the construction and real estate industry as well as other interested readers who would like to learn the theory and practice of VM.

Chapter 1 introduces the concept and methodology of Value Management. Chapter 2 presents the process which is known as the Job Plan for Value Management. Chapters 3 and 4 lay the theoretical foundations for group dynamics and group facilitation that is applied in Value Management respectively. Chapters 5 and 6 are the implementation and application of VM respectively. Chapter 7 explains the use Group Support Systems, a branch of IT, which can be applied to improve VM studies. Chapter 8 illustrates the application of Value Management in the briefing process of construction projects. Chapter 9 focuses on the measurement of performance for the Value Management studies. Chapter 10 provides several case studies which applied VM in the development process in construction.

Feedback and comments are welcome for future editions of this book.

Foreword

To promote enhancement of the surveying practice and the construction industry at large, the Hong Kong Institute of Surveyors ('HKIS') sponsored writing of this book entitled 'Value Management in Construction and Real Estate: Methodology and Applications'. Construction projects, nowadays, have been increased in scale and complexity. Clients demand their projects to be built in a quality, economical, efficient and safe way and achieve value for money. Consequently, construction professionals are obliged to advance ourselves to meet satisfaction of their Clients. Value Management has continued to play an important role for construction projects, in both public and private sector, to bring value and satisfaction to the clients and ideally the ultimate users.

This book is a valuable resource for those who want to know more about the full ranges of issues concerning value management and to carry out value management exercise successfully. The book is a great reference for academics and practitioners as well as students. This book is written similar to a hand-book style which has rendered a very handy and concise reference to readers. The fundamental concepts and applications of value management are illustrated by diagrams and case studies. The first six chapters introduce the development and methodology of value management. The subsequent three chapters reveal the research findings of the authors in the past ten years. The last chapter presents the value management case studies in Hong Kong.

It is my pleasure to highly recommend this book to those which are both experienced and inexperienced with value management studies. The book is not only easy to read for beginners but also enlightening for more experienced readers.

Sr Vincent Ho
President 2014-15
The Hong Kong Institute of Surveyors
March 2015

1 The VM Methodology

*Geoffrey Q.P. Shen, Ann T.W. Yu,
and Jacky K.H. Chung*

1.1 Introduction

This chapter introduces the background and methodology of Value Management (VM). It provides definitions and the historical development of VM, discusses the rationale behind VM, and identifies the key components of VM.

1.2 Learning Objectives

Upon completion of this chapter, you should be able to:

1. Provide a definition of VM
2. Summarise the historical development of VM
3. Explain the methodology of VM
4. Identify the key components of VM

1.3 Definitions

> **What is your definition of VM? Please write it down and compare it with the definitions given below.**

Value analysis is a problem-solving system implemented by the use of a specific set of techniques, a body of knowledge, and a group of learned skills. It is an organised creative approach that has for its purpose the efficient identification of unnecessary cost, i.e. cost that provides neither quality nor use nor life nor appearance nor customer features.

Miles (1972)

Value engineering is a creative, organised approach whose objective is to optimize cost and/or performance of a facility or system.

Dell'Isola (1982)

Value Engineering (synonymous with terms value management and value analysis) is a function-oriented, systematic team approach to provide value in a product, system, or service. Often, this improvement is focused on cost reduction, however, other improvement such as customer perceived quality and performance are also paramount in the value equation.

SAVE (1998)

VA is a management technique which analyses, by means of a systematic approach, how to reduce cost whilst taking into account customer requirements; it not only assesses the degree of innovation desired or allowed for in the product or service, but also covers the implementation and follow-up of solutions proposed and therefore strengthens companies' innovative capacity and competitiveness.

Commission of the European Community (1991)

Value management is a structured and analytical process which seeks to achieve value for money by providing all the necessary functions at the lowest cost consistent with required levels of quality and performance. [This definition uses the term value management synonymously with the terms value analysis and value engineering].

AS/NZS 4183:1994

Value Management is a service which maximizes the functional value of a project by managing its evolution and development from concept to completion, through the comparison and audit of all decisions against a value system determined by the client or customer.

Kelly and Male (1993)

Value Management is an organised function-oriented systematic team approach directed at analysing the functions and costs of a system, supply, equipment, service or facility, for the purpose of enhancing its value, through achieving the required functions specified by the clients at the lowest possible overall cost, consistent with requirements for performance.

Shen (1993)

1.4   Historical Development

Lawrence D. Miles of the General Electric Company, USA, founded the VM methodology after the Second World War. The major events in the historical development of the VM methodology are summarised in Table 1.1.

Table 1.1 VM History and Major Developments

Year	Major Historical Developments
1947	Function Analysis and the VM job plan were first developed by Lawrence D. Miles of the General Electric Company (GEC) in the USA.
1947	Function Analysis was developed in the practice of work study by Imperial Chemical Industries Plc. in the UK.
1954	US Department of Defence adopted VM when the Navy's Bureau of Ships set up a formal VM programme for cost improvement during design. They called it 'value engineering'.
1959	Society of American Value Engineers (SAVE) was established.
1962	USA's Ministry of Defence set up VM programmes in large scale bidding procedures.
1963	Dell'Isola first applied VM to buildings by introducing Value Engineering to the US Navy's Facilities Engineering Command. The US General Service Administration began to use VM shortly thereafter.
1965	Function Analysis System Technique (FAST) was introduced by Charles Bytheway of the UNIVAC Division of Sperry Rand Corporation, at the 5th SAVE National Conference.
1966	The Value Engineering Association was established in the UK. The present name of the society is the Institute of Value Management.
1969	The National Aeronautics and Space Administration began formal VM studies and training.
1970	The US congress endorsed contractor incentive clauses for the Department of Transportation (the Federal Register, Vol. 34, No. 173, 1970).
1972	SAVE 12th annual conference emphasized the application of VM in the construction industry.
1975	The US Environmental Protection Agency mandated that VM be used during the design of all wastewater treatment facilities over $10 million.
1988	RICS published 'A Study of Value Management and Quantity Surveying Practice', by Kelly and Male, which demonstrated the potential of the QS profession using VM in the UK. A second report, 'The practice of Value Management: Enhancing Value or Cutting Cost', was published in 1991.
1990	Under the funding of the European Community's Strategic Programme for Innovation and Technology Transfer (SPRINT), a number of research projects into VM were launched in several countries of the Community.
1993	US H.R.133 'Systematic Application of Value Engineering Act of 1993' requires US Federal agencies to apply value engineering to programmes, projects, systems and products comprising 80% of an agency's budget.
1993	US H.R.2014 'Value Engineering Better Transportation Act of 1993' was introduced. This innovative bill provides incentives to US State Transportation agencies to conduct VE by increasing the percentage of Federal grant money based on performance and savings.
1994	Introduction of Australian/New Zealand Standard on Value Management.
1994	The Central Unit of Purchasing of H.M. Treasury in the UK commissioned Davis Langdon Consultancy to draw up recommendations for a VE Guide for project sponsors.
1995	The Hong Kong Institute of Value Management was established.

Table 1.1 Continued.

Year	Major Historical Developments
1998	Works Bureau Technical Circular 16/98 and Planning Environment and Land Bureau Technical Circular 9/98 'Implementation of Value Management' called for VM studies in all major public sector projects in Hong Kong.
1999	First European Standard on Value Management 'EN 12973' was approved, including minimum training requirements for value management practitioners.
2001	Hong Kong's Construction Industry Review Committee recommended the use of VM.
2002	Hong Kong's Environment, Transportation, and Works Bureau released technical circular 35/2002, calling for wider use of VM in the public sector.
2002	Hong Kong's Works Bureau recommends VM for projects over HK$200 million.
2003	Hong Kong's Transport Department and the Correctional Services introduce VM.
2003	The Professional Services Development Scheme of the HKSAR approves the 'Enhancement of Construction Value Management Professionalism for the New Generation' scheme to update the VM knowledge and skills of the construction industry professionals.
2006	The Professional Services Development Scheme of the HKSAR approves the 'Improving Value Management (VM) [...]' scheme to help improve the Value Management practice in Hong Kong.
2007	The Institute of Value Management Australia (IVMA) contributes to the rewriting of the Australian/New Zealand Standard on Value Management which is updated and re-released. (AS 4183-2007).
2007	SAVE release their first 'Value Methodology Standard and Body of Knowledge' booklet with information and guidelines on VM.
2009	SAVE celebrate their 50th anniversary.
2009	SAVE and the Hong Kong Institute of Value Management sign a mutual support agreement on VM related activities.
2010	SAVE International's 50th annual conference held in California. Governor A. Schwarzenegger declares the week of the conference for 'Value Engineering Week for California'.
2012	European Standard on Value Management 'EN 16271' is approved and extends upon 'EN12973'.
2013	The Hong Kong Institute of Value Management advises the Development Bureau of the HKSAR to revise and update their 1998 and 2002 circulars to reflect the advances and applicability of VM.
2014	Value Engineering is a prominent feature of the US General Services Administration (GSA), responsible for the US Governments Real Estate function.
2015	The Hong Kong Institute of Value Management celebrates their 20th anniversary which coincides with hosting their 12th international VM conference.

1.5 VM Methodology

Rationale of VM

The rationale behind the use of VM is that there are always elements in any project that contribute to poor value and that they need to be addressed. Norton and McElligott (1995) identified these elements as:

- Not enough time to do the job
- Habitual thinking
- Poor communications
- Lack of coordination between designer and operations personnel
- Outdated standards or specifications
- Absence of state of the art technology
- Honest false beliefs
- Prejudicial thinking
- Lack of needed experts
- Unnecessarily restrictive design criteria
- Scope of changes for missing items
- Lack of needed information

In each element there is an opportunity to improve value for money. VM suggests that project performance improves if unnecessary costs have been identified and removed successfully. In order to achieve this efficiently and effectively, a methodology called the 'VM Job Plan' has been developed to standardise the process. In general, value for money can be improved in the following ways (EUR 16096 EN, 1995):

- Provide for all the required project functions at a reduced cost
- Provide additional desirable project functions without adding to the cost
- Provide additional desirable project functions while at the same time reducing costs

---

Do you agree that project performance improves if poor value elements are identified and removed in a construction project?

VM and Cost Reduction Techniques

Most people in the construction industry apply the term VM to any methodology used to seek ways to reduce costs. While VM does usually result in cost reductions, it is very different from all other cost reduction techniques applied to construction projects. There is also a common misconception that VM is something designers routinely do during the design development process.

Norton and McElligott (1995) suggested that VM is a 'systematic, multi-disciplinary effort directed toward analysing the functions of projects for the purpose of achieving the best value at the lowest life cycle cost' and summarised the major difference between VM and cost reductions techniques as shown in Table 1.2.

Table 1.2 Major Differences between VM and Cost Reduction Techniques

Elements	Description
Systematic process	The job plan is applied in VM. It is a formal and structured process that has a definite beginning and a definite end.
Multidisciplinary approach	VM involves bringing a group of individuals together who collectively and authoritatively analyse all aspects of the project as a team.
Functions	VM involves analysing the functions of a project, which distinguishes VM from other design review or cost reduction activities.
Value	VM studies aim to improve value rather than to only reduce costs.

Adapted from Norton and McElligott, 1995.

1.6   Key Components

As shown in Figure 1.1, the following five components essential to the success and advancement of VM methodology have been identified: Functional Approach, Function Cost Approach, Environment for Creativity, Organised Team Approach, and Organised Job Plan.

Functional Approach

The Functional Approach is an essential element of the VM methodology with a relatively long history of evolutionary development (Gregory, 1984). It consists of a group of techniques that differentiate it from traditional cost reduction and cost planning efforts. The objective of the Functional approach is to forget the product as it exists and concentrate only on its necessary functions as required by the client. This approach leads to a systematic identification and clear definition of the client requirements, an improved functional understanding of the design problem, and an effective accomplishment of those functions.

Function Cost Approach

Value = Worth / Cost

When looking at a function in VM, worth is defined as the lowest cost required to perform the bare minimum of the function while disregarding the criteria or codes associated with it. By comparing the actual costs of a function to its worth, it is

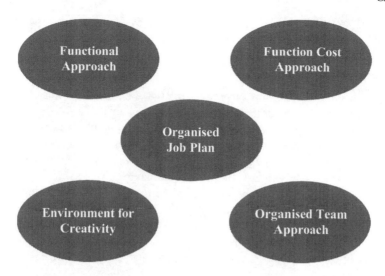

Figure 1.1 Key Components of VM

possible to get an indication of the potential value improvement that can be achieved. This technique can help the VM study team to decide which areas to focus on.

Environment for Creativity

The creative stage is one of the most exhilarating stages in the VM process. A detailed discussion about creativity is provided in Chapter 4.

Organised Team Approach

The term 'organised and systematic approach' is commonly referred to as the job plan for a workshop, which is greatly emphasized in VM methodology. The plan shows the systematic procedures for accomplishing all the necessary tasks associated with a VM study. It is at the centre of all existing European VM Standards, particularly the French and German. There are a variety of VM job plans, each with a number of phases ranging from five to eight.

Organised Job Plan

The most commonly used organised job plans are:

- Charette Job Plan
- SAVE 40-hour Job Plan
- Contractor's Change Proposal

- Truncated Workshop
- Concurrent Study

Charette Job Plan

The Charette Job Plan attempts to rationalise the client's brief primarily through functional analysis of space requirements. If time is available for the study, this plan could be broadened to include other issues concerning the client requirements. The main focus is to ensure that the designers fully understand the client requirements. The biggest advantage of this job plan, as Kelly and Male (1991) explained, is that it is considered by many clients to be an inexpensive and effective method of briefing the design team and clarifying their own requirements in the least possible time; it takes less than two days compared with five days required by the 40-hour Job Plan. As the study is usually carried out in the very early design stage of a project, it not only avoids potentially abortive design work but also plays a major role in controlling costs and enhancing the value of a project.

SAVE 40-hour Job Plan

The SAVE 40-hour Job Plan is the most commonly accepted formal approach and is summarised in Table 1.3.

Contractor's Change Proposal

This common form of delivering VM to a construction project, if well managed, can produce excellent results for minimum effort. This statement has been proved over the past four decades by many VM practitioners, and this job plan is regarded as a VM milestone by SAVE and many other VM organisations.

Truncated Workshops

When the estimated project value is lower than the US$2-3 million considered as the lower limit for a full 40-hour workshop, the 10 percent rule of thumb is used to determine how much can be spent on the workshop. For example, a project of $1 million would have a target saving of $100,000 and a workshop fee of $10,000, based on which the duration and the number of professionals involved can be worked out.

Concurrent Study

With this approach, the design team holds regular meetings under the supervision of the value engineer to review the design decisions that have been taken. This method is lauded for answering much of the criticism usually expressed towards the standard 40-hour study. To determine the fee for the value engineer, the extent of his contributions and involvement need to be established in advance. Likewise,

Table 1.3 The SAVE 40-hour Job Plan

Phase	Description
Information Phase	The main tasks in this phase include collecting historical cost data, the client's requirements, design standards, and specifications in order to obtain a thorough understanding of the project.
Function Analysis Phase	This identifies and defines significant functions of the project and where necessary determines the cost of each function, and, in some cases, identifies who is responsible for performing each function.
Creativity Phase	The main task in this phase is to generate numerous alternatives for accomplishing basic functions required by the client. This is done by means of creative stimulating techniques, such as Brainstorming, Synectics, Morphological Chart, and Lateral Thinking. This phase answers the question of what else would work to perform the basic functions.
Evaluation Phase	This phase establishes a number of criteria by which to evaluate and select alternatives generated during the creativity phase. Various models can be used, such as cost models and energy models, and it answers the question of what the alternatives would cost.
Development Phase (or Proposal Phase)	This phase investigates selected alternatives in sufficient depth to develop them into written recommendations suitable for implementation. This involves not only detailed technical and economic evaluation but also consideration of the probability of successful implementation.
Presentation Phase	This phase involves defining and quantifying results on order to prepare and present a Value Management Change Proposal (VMCP) for the final decision makers. The proposal usually includes an analysis of the potential benefits and a statement of the follow-up procedures necessary to ensure implementation.
Implementation Phase	This phase ensures that the recommendations in the VMCP are fully implemented, which includes providing assistance and clearing up misconceptions, auditing and resolving problems that may develop in the implementation process, and comparing the actual results with what were originally expected by the VM study team.

Adapted from SAVE, 1998.

the members of the design team involved in the fee-bid will have to account for the extra involvement required. In construction management contracts, the concurrent study is also suitable because the design is carried out in phases simultaneously with the construction (fast-tracking).

One of the disadvantage of the concurrent study is in proving actual value increases at the time of implementation, as some will find it easy to argue that the budget was set too high initially (Kelly *et al.*, 2004).

Drawbacks of Organised Job Plans

The following are some drawbacks of organised job plans:

- Firstly, it is difficult to assemble the key project participants for such a concentrated period and retain their undivided attention.
- Secondly, since much of the session must be devoted to educating participants who are not familiar with the VM processes, it is rather difficult to bring the processes to bear on the problem at hand.
- Thirdly, the evaluation and development are particularly difficult to complete effectively in such a short time, because many ideas proposed in the speculative phase often require intensive design and engineering analysis, particularly where they involve long-term life cycle cost trade-offs.
- Finally, as the VM study is often isolated from project cost management, it is not unusual for a successful VM workshop to be followed by cost overruns because of this lack of integration.

One way to solve these problems is to disperse the VM processes throughout the project from inception to completion, including feasibility study, project definition, concept design, design development, contract documentation, procurement and construction, hand-over and operation, and feedback and evaluation.

Although this procedure has some advantages, including a flexible timetable for those participating in the process, the team effect of the original SAVE 40-hour Job Plan is virtually eliminated. The real challenge is that to implement this proposal would require each VM team member to have advanced computers with the necessary software to provide them with the latest project information in order to make comments or evaluations.

Kelly and Male (1991) pointed out that, according to VM theories, the proportion of time allocated to each phase of the 40-hour workshop is unbalanced. The information assimilation, development, and presentation phases are too long, while the time allocated to functional analysis is not long enough. This shortcoming can be partly overcome by developing a knowledge-based system, equipped with VM domain knowledge, that could reduce the time allocated to tasks such as information retrieval, alternative evaluations, and presentations, thereby allowing more time to be allocated to the all-important function analysis phase.

1.7 References

Australian/New Zealand Standard (1994). *Value management*. AS/NZS 4183, Standards Australia, Standards New Zealand.

Commission of the European Communities (1991). *Value analysis glossary*, European Community Strategic Programme for Innovation and Technology Transfer (SPRINT). Luxembourg: Report EUR 13774 EN, DG XIII, L-2920.

Dell'Isola, A.J. (1982). *Value engineering in the construction industry*, 3rd edn. New York: Van Nostrand Reinhold Company Inc.

Gregory, S.A. (1984). Design technology transfer, *Design Studies*, 5(4), 203–218.

Kelly, J. and Male, S. (1993). *Value management in design and construction: the economic management of projects*. London: E & FN Spon.

Kelly, J.R. and Male, S.P. (1991). *The practice of value management: enhancing value or cutting cost?* London: Royal Institution of Chartered Surveyors Publications.

Kelly, J., Male, S., and Graham, D. (2004). *Value management of construction projects*. Oxford: Blackwell Science Ltd.

Miles, L.D. (1972). *Techniques of Value Analysis and Engineering*, 2nd edn. London: McGraw-Hill Publishing Company Ltd.

Norton, B.R. and McElligott, W.C. (1995). *Value Management in Construction: A Practical Guide*. Basingstoke: Macmillan Press.

SAVE International (1998). *Value Methodology Standard*, 3rd edn.

Shen, Q.P. (1993). *A knowledge-based structure for implementing value management in the design of office buildings,* thesis submitted to the University of Salford for the degree of doctor of philosophy. UK: British Library Document Supply Centre.

2 The VM Process

Geoffrey Q.P. Shen, Ann T.W. Yu,
and Jacky K.H. Chung

2.1 Introduction

By considering the most popular VM job plan in the construction industry, the SAVE 40-hour Job Plan, this chapter explains the objectives, processes, and techniques of the VM process, as well as the information, function analysis, creativity, evaluation, development, and presentation phases of a VM workshop.

2.2 Learning Objectives

Upon completion of this chapter, you should be able to:

1. Explain the objectives, processes, and techniques of the VM process
2. Describe the information, function analysis, creativity, evaluation, development, and presentation phases of a VM workshop

2.3 The VM Job Plan

The VM job plan is a sequential approach to implementing the core elements of value management. It outlines specific steps to effectively analyse a product or service, and develops the maximum number of alternatives to achieve the functions required for the product or service (SAVE, 1998a). Dell'Isola (1997) suggested that the job plan is an organised problem-solving approach, which distinguishes VM from other cost-cutting exercises. There are a variety of VM job plans, such as Charette Job Plan, SAVE 40-hour Job Plan, VM Audit, Contractor's Change Proposal, Truncated Workshop, and Concurrent Study. This chapter considers the SAVE 40-hour Job Plan, which is the most popular approach in the construction industry. According to the Value Methodology Standard (SAVE, 1998a), the SAVE 40-hour Job Plan comprises the following three major stages: (i) pre-workshop stage, (ii) workshop stage, and (iii) post-workshop stage. A summary of the stages is provided in Table 2.1.

Table 2.1 The VM Job Plan

Stage/Phase		Main Tasks
PRE-WORKSHOP STAGE		Collect User/Customer Attitudes
		Complete Data File
		Determine Evaluation Factors
		Define Scope of the Study
		Build Data Models
		Determine Team Composition
WORKSHOP STAGE	Information Phase	Complete Data Package
		Modify Scope
	Function Analysis Phase	Identify Functions
		Classify Functions
		Develop Function Relationships/Models
		Establish Function Worth
	Creative Phase	Create Quantity of Ideas by Function
	Evaluation Phase	Rank and Rate Alternative Ideas
		Select Ideas for Development
	Development Phase	Conduct Benefit Analysis
		Complete Technical Data Package
		Create Action Plan
		Prepare Final Proposals
	Presentation Phase	Present Oral Report
		Prepare Written Report
		Obtain Commitments for Implementation
POST-WORKSHOP STAGE		Complete Changes
		Implement Changes
		Monitor Status

Adapted from SAVE, 1998a.

2.4 Pre-Workshop Stage

The pre-workshop stage (sometimes referred to as the pre-study stage) provides an opportunity for all parties to understand project issues and constraints, as well as to give and receive information before the VM workshops. The preparation tasks involve the following six areas:

- Collecting user/customer attitudes
- Completing the data file
- Determining evaluation factors
- Defining the scope of the study
- Building data models
- Determining team composition

The pre-workshop stage determines the team structure and defines the study. Clients can employ VM facilitators to organise and facilitate briefings for them. As stakeholders come from different organisations that have different objectives and requirements, the composition of team members is critical. Clients are advised to select stakeholders carefully as this may directly influence the project outcome. The construction process, which plays such a key role in the project delivery cycle, is usually neglected at the design stage of a traditionally procured contract. However, it is strongly recommended to include contractors in the VM study team, which should comprise the following members:

- Project managers
- Client representatives
- Design team members
- Contractors
- End-users
- Facilitators (specialist VM facilitator or trained project manager)

Facilitators also assist clients to define the scope and objectives of the study, as well as ensuring that sufficient information is made available to all members of the team (SAVE, 1998a).

> **How would you arrange the pre-workshop phase in a one-day workshop?**

2.5   Workshop Stage

The workshop stage comprises the following six major phases:

1. Information Phase
2. Function Analysis Phase
3. Creativity Phase
4. Evaluation Phase
5. Development Phase
6. Presentation Phase

Information Phase

The objective of the information phase is to compile data in order to produce an information base for the VM study, finalise the study's objectives, clarify assumptions, and review the scope of the study. This phase ensures that all members of the team fully understand the background, objectives and constraints of the study so as to broaden their perspectives beyond their particular area of expertise. An introductory presentation by a representative of the client is usually followed by a description of objectives for the project, after which the stakeholders take it in turn to express their views on the project's requirements and constraints. Although it is inevitable that conflicting views will be expressed, consensus is not likely to be reached until after the function analysis phase. A summary of the information phase is provided in Table 2.2.

Table 2.2 Information Phase

Typical Objective	Typical Questions	Typical Activities/ Techniques
Complete Data Package	What is VM purpose?	Presentation
Modify Scope	What is rationale?	Graphics
	What is timetable?	Cost, energy, area models
		Pre-reading
		Life cycle cost

Function Analysis Phase

Function analysis is the heart of VM methodology such that it is the primary activity that distinguishes VM from all other improvement practices. The objective of this phase is to identify, classify, and develop the most beneficial functions for further study. A summary of the function analysis phase is provided in Table 2.3.

Table 2.3 Function Analysis Phase

Typical Objective	Typical Questions	Typical Activities/ Techniques
Identify Functions	What does it do?	Function Analysis
Classify Functions	What does it cost?	FAST diagram
Develop Function	What must it do?	Function hierarchy
Relationships	What are the performance criteria?	Priority Matrix
Establish Functions' Worth	What are the quality criteria?	

(i) Identifying Functions

A function of a product is identified by its specific purpose or intended use. This definition focuses on the characteristics that make the product work, sell, produce revenue, or meet the requirements for a specific purpose (Dell'Isola, 1982).

An active verb and a measurable noun expresses a function. The verb answers the question, 'What is it to do?' The noun answers the question, 'What does it do it to?' It is important that the verb is active and thus describes what the item does. Using a general verb such as 'provide' can seem straightforward, but since this might imply a solution it should be avoided. The noun should be quantifiable or measureable for 'use-type' functions and should describe what the verb is acting upon (Norton and McElligott, 1995). Some examples of identifying functions are: absorb energy, collect heat, support weight, and incorporate space.

This two-word simplification isolates the function and keeps it simple without excessive information that could otherwise force the designer to try to decide what data is fundamental and should be retained, and what is unimportant and should be rejected. While helping the designer it also helps ensure full understanding by all team members irrespective of their knowledge, educational and technical backgrounds (SAVE, 1998a). Using a commercial building project as an example, the most obvious basic function of the project could be to provide office space. However, this may not necessarily be the real intention, which might be to provide a 'world class' commercial centre or a new focal point for the entire commercial area. At the project level, the functions of a world-class commercial centre can be described as to 'satisfy users' and 'attract users'. McGeorge and Palmer (1997) suggested that the function of an item should only be defined in terms of a verb and a noun if it is not fully understand. It should ideally be identified by what is to be accomplished by a solution, not how it is to be accomplished.

(ii) Classifying Functions

BASIC AND SECONDARY FUNCTIONS

Once the functions have been identified, they are categorised as either a basic function or a secondary function in order to show their relative importance.

The primary purpose(s) that an item or service is designed to accomplish when operating in its normal environment, is referred to as a basic function. Once a basic function has been defined, it cannot change and a loss in its functionality causes a loss in the value of the project. The basic function is essential to meet the purpose of the product, structure or service, and must be accomplished (SAVE, 1998a). Crow (2001) explained that functions designated as basic will not change, but the way those functions are implemented is open to innovative speculation. For project-level function analysis, basic functions are those essential to satisfy the needs and requirements of the client and answer the question 'what must it do?' There may be more than one basic function of a project. For example, the functions of 'satisfy users' and 'attract users' are considered as basic functions of a 'world class' commercial centre.

A secondary function supports the basic function. It is the direct result of a design decision that was taken to achieve the purpose of the basic function (SAVE, 1998a). Secondary functions are auxiliary features that are not essential in achieving a basic function. These functions describe what happens beyond the basic functions, without directly influencing it, and they usually occur as a result of the method chosen to achieve a basic function (Norton and McElligott, 1995). For example, the functions of 'ensure safety' and 'ensure comfort' may be considered as secondary functions of a 'world class' commercial centre.

> **Can you list the basic functions and secondary functions of a classroom?**

The distinction between basic functions (what is required in the project and provides value) and secondary functions (what is not required and has no value) is vital to the success of identifying client requirements. Norton and McElligott (1995) pointed out that, while it can be difficult and time consuming, the function classification forces team members to have a deeper level of understanding of the project and its constituent parts. It also eliminates the chance that the proposed project design is contrary to the project's requirements and needs. In order to classify functions efficiently, the most common approach is to list the physical parts of the project, or steps of a procedure, and attempt to define the functions associated with each part or step. For larger building projects, each requirement of the client, spaces of a building, elements of a structure, etc. may be the appropriate level of analysis (SAVE, 1998a).

USE AND AESTHETIC FUNCTIONS

Functions can also be classified into use functions and aesthetic functions.

- Use functions involve an action that clients want performed, whereas
- Aesthetic functions please the facility's clients

For example, a client may want space provided, an improved environment, security ensured (use functions), as well as specific colour, shape and appearance to appeal to his/her staff and customers (aesthetic functions). Table 2.4 shows the verbs and nouns that are commonly used for describing use functions and aesthetic functions.

Table 2.4 Verbs and Nouns for Use and Aesthetic Functions

	Use Functions	*Aesthetic Functions*
VERB	Absorb, change, circulate, collect, condition, conduct, connect, contain, control, convey, create, detect, distribute, enclose, exclude, improve, insulate, protect, reduce, resist, support, ventilate	create, ensure, establish, experience, feel, finish, improve, increase, reflect, satisfy, smell, taste, think
NOUN	air, compression, current, elements, energy, fire, flow, fluids, force, heat, landscape, load, materials, objects, oxidation, parking, people, power, radiation, sheer, sound, space, temperature, tension, voltage, weight	appearance, balance, beauty, colour, convenience, features, feeling, image, prestige, preparation, shape, space, style

(iii) Developing Function Relationships

The task of developing function relationships involves establishing the linkages among functions by using the Functional Analysis System Technique (FAST). FAST is one of the most popular techniques in function analysis. This technique is based on the intuitive logic of HOW-WHY relationships and graphically displays them in a diagrammatic form. A horizontal chart enables functions within a project to be displayed in a logical sequence and their dependency tested rigorously according to the following rules (SAVE, 1998b):

- The sequence of functions on the critical path proceeding from left to right answers the question 'how is the function to its immediate left performed?'
- The sequence of functions on the critical path proceeding from right to left answers the question 'why is the next function performed?'
- Functions occurring at the same time or caused by functions on the critical path appear vertically below the critical path.
- The basic function of the study is always farthest to the left of the diagram; to the left of all functions within the scope of the study.

A diagram of a 'world class' commercial building is shown in Figure 2.1, which illustrates the concept and mechanics of the FAST approach. A FAST diagram has a horizontal directional orientation described as the HOW-WHY dimension. Starting with the highest order function (the ultimate goal of the project) on the left, and working across to the right, we ask the question HOW. Operating in the opposite direction we ask the question WHY, which checks whether the logic of the diagram is correct. The desired result (a 'world class' commercial centre) is the highest-order function and is placed on the left side of the diagram. Following in priority, the basic functions representing the purpose of the design are then placed next to the highest-order function. In the example there are multiple basic functions, such as 'satisfy users', 'assure efficiency', 'attract users', etc. The basic functions are then broken down into secondary functions answering the 'how can the basic function be achieved' question. For example, the basic function of 'satisfying users' can be achieved by 'ensuring safety' and 'providing services', etc. If the diagram is read from right to left it can answer the question of WHY a particular function is necessary, which should help verify the structure and validity of the diagram (McGeorge and Palmer, 1997). Secondary functions can then be further developed into lower-level functions to explain how individual functions can be achieved.

Figure 2.2 shows that the function 'ensure comfort' can be further broken down to lower-level functions such as 'regulate temperature' or 'control noise'. The extent to which the levels of functions should be included in a FAST diagram depends on the client, the level of design completed, and the project generally (McGeorge and Palmer, 1997). The following summarises what a FAST diagram is and is not:

- It is an effective management tool to display functions in a logical sequence and to rigorously test their dependency.

- It enables the drawing of relationships between the individual functions that need to be performed and the object of analysis, and then shows those relationships on a chart (Akiyama, 1991).
- It enables an appreciation of problems from a broad perspective, and creates an understanding of the function relationships on larger scope projects.
- It permits people with different technical backgrounds to communicate effectively and resolve issues that require multi-disciplined considerations (Norton and McElligott, 1995).
- It is not a panacea for project problems. It displays functions in a logical sequence, internally prioritising and testing their dependencies, but it does not decide how a function should be performed (specification), when (not time oriented), by whom, or how to perform the function.

(iv) Assigning Weightings to Functions

To prioritise between functions and show their relative importance, a weighting is assigned to them. There are different methods to show the importance. One such approach is to distribute 100 points to the first level functions, then for the second level functions, the points are distributed according to their corresponding first level functions and their relative importance. This ensures an easier comparison between functions at the same level. Systemising the importance of functions helps clients to understand their requirements and needs better, and the designers get useful information in their quest to satisfy the client. The relative importance of the functions for a 'world class' commercial centre is illustrated in Figure 2.1. The cost factor can also be incorporated with the weightings of functions so that a function-cost relationship is developed. An estimated cost is assigned to each function in order to illustrate its financial implications in the briefing process. However, it could be very difficult to allocate costs to functions because functions are not related to design in the process.

(v) Assigning Flexibility to Function

Assigning flexibility to function, which is proposed in a Functional Performance Specification (FPS), is an additional step built into the SAVE 40-hour Job Plan to further investigates the functions identified in the analysis phase. The FPS is a document in which an enquirer expresses needs in terms of user-related functions (service functions) and constraints (EUR16096 EN, 1995). The FPS technique is an additional tool in the arsenal of the usual VM exercise. The FPS is applied after the functions have been identified, defined and weighted. The FPS is an information tool that provides a high level of detail for each of the functions in a construction project, this helps all concerned understand the requirements of the client more clearly (Masson, 2001). To achieve this, flexibility is assigned to each function, following the steps illustrated in Table 2.5.

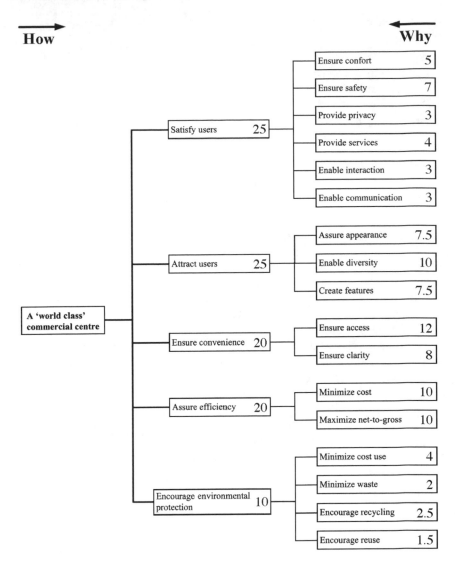

Figure 2.1 A FAST Diagram of a 'World Class' Commercial Centre

Adapted from Shen *et al.*, 2004.

(A) DEFINING CRITERIA

The first step is to define all the criteria that will be used to evaluate and measure if a function is achieved. Generally, there might be several criteria for each function. Referring to the lower-level functions illustrated in Figure 2.2, the criteria for the function of 'regulate air quality' are defined as 'efficient circulation', 'minimal airborne contaminants' and 'well located air inlets'.

(B) DEFINING LEVELS OF CRITERIA

The second step is to define the levels of each function-criterion that are acceptable to satisfy the need. For example, avoiding close proximity of outdoor intake to sources such as garages, loading docks, building exhausts, and outside construction projects is considered an acceptable level for the criteria of 'well-located air inlets'.

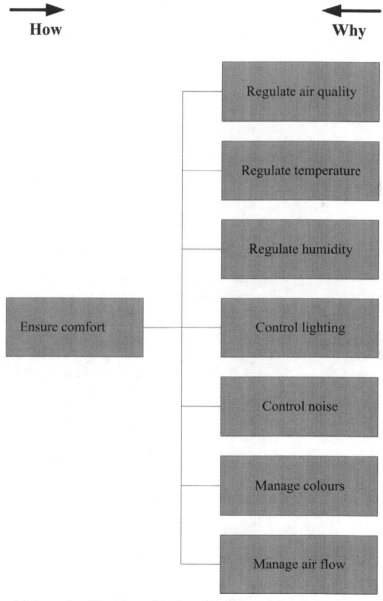

Figure 2.2 Lower-level Functions of the Function of 'Ensure Comfort'

Adapted from Shen *et al.*, 2004.

Table 2.5 Example of a FPS

Functions	Criteria	Levels	Flexibility
Regulate Air Quality	Well-located air inlets	Avoid close proximity of outdoor intake to sources such as garages, loading docks, building exhausts, outside construction projects.	F1
	Efficient circulation	Minimum ventilation (i.e. the introduction of fresh air to replace stale air): • about 0.5 to 3 air changes/hour depending upon density of occupants. • values per occupant range from 5 to 25 litres/sec/person (Baker and Steemers, 2000).	F2
		Air movement to cool heat sources (people, lighting, equipment): • average air velocity during winter not to exceed 30 feet/minute1 (fpm). • average air velocity during summer not to exceed 50 fpm.	F2
	Minimal airborne contaminants	High efficiency filter to be used for HVAC* system to remove bacteria, pollen, insects, soot, dust, and dirt (ASHRAE** dust spot rating of 85% to 95%) (EPA, 2001).	F1
		Areas from which fumes need to be extracted (such as toilet facilities, cooking facilities, and parking garages) must be maintained at a lower overall pressure than surrounding areas. These areas should also be isolated (disconnected) from the return air system so that contaminants are not transported to other parts of the building.	F0
	Allow for individual control	Local control system to modulate airflow.	F0
		Control switches to be conveniently located and properly instructed.	F0

F0: The criteria is an absolute must, not negotiable, all effort must be made to meet this level, whatever the cost.

F1: The criteria is a must if at all possible, no discussion unless there is a very good reason.

F2: The criteria is negotiable, hope this level is reached, ready to discuss.

F3: The criteria is very flexible, this level is proposed but is open to any suggestion.

(C) ASSIGNING FLEXIBILITY TO CRITERIA

The third step is to assign flexibility to each criterion, using a four-scale assessment system, in order to examine its negotiability.

The FPS requires all the criteria by which each function will be measured, the expected level of satisfaction for each criteria level, and the corresponding flexibility allowed (Masson, 2001). A solely functional approach eliminates all restrictions and shows a precise definition of the requirements that are essential to the products or services. The involved team can then contribute with the best of their expertise and creativity to formulate a proposal (EUR16096 EN, 1995).

> **Do you know the differences, including function and purpose, between the FAST and the FPS?**

Creativity Phase

The objective of the creativity phase (sometimes called the speculation phase) is to come up with a large quantity of ideas for performing each of the functions selected for study. It is important that this is a creative effort with no judgement on ideas or discussions constrained by habit, tradition, negative attitudes, assumed restrictions or specific criteria. The actual quality of the ideas generated here, is developed in the next phase (SAVE, 1998a).

Table 2.6 Creativity Phase

Typical Objective	Typical Questions	Typical Activities/Techniques
Create Quantity of Ideas by Function	What are the alternatives? What else satisfies function?	Brainstorming Golden technique Synectics Lateral thinking

> **What are the reasons for producing a large quantity of ideas in the creativity phase and how do you achieve this in a VM workshop?**

There are two keys to successful creativity. The first is to develop ways to perform the selected functions rather than to waste time and energy on conceiving of ways to design a product or service. The second key is to combine past experience to form solutions that will fulfil the requirements of the functions at a lower cost and with improved performance (SAVE, 1998a). However, there are a number of blocks to creativity as shown in Table 2.7.

There are numerous well-accepted idea-generating techniques, including brainstorming, synectics, and the plus-minus-interesting (PMI) technique, details of which are discussed in Chapter 3.

The Brainstorming Process

Brainstorming is the most popular technique in the creative phase. It requires that members of a group consider a function and contribute any suggestion which will

Table 2.7 Blocks to Creativity

	Blocks to Creativity
Perceptual Blocks	Individuals perceive things in different ways, and therefore we often block out information which conflicts with our perceptions. Construction professionals, including clients, architects, engineers, surveyors and other parties, all see a project from their own particular perspective, which may be rather narrow. It is important for each discipline to be empathetic to other disciplines' perspectives so that all members get a better picture of the problems as a whole.
Habitual Blocks	Most of us are guilty to some degree of following procedures unquestionably just because we have always done things that way in the past. This may derive from unchallenged standard specifications and briefs or habitual design practices in construction.
Emotional Blocks	An upbringing of derision for mistakes and failure makes us all averse to putting forward ideas that may be sub-optimal or incorrect. This aversion or fear suppresses the seedlings of ideas that may spawn new and valuable approaches.
Cultural and Environmental Blocks	The culture and environment in which we are raised impacts on our perceptions. The cultural aspects can impose a strong desire for conformity which obviously stifles creativity.
Professional Blocks	Professional regulation and education tends to confine us within boundaries of behaviour and perception. While such boundaries are useful in certain contexts, their existence should be borne in mind so that they do not inhibit creative solutions.

Adapted from Norton and McElligott, 1995.

expand, clarify or answer that function. Research has indicated that the number of good suggestions remains fairly constant as a proportion of wild suggestions, so the more suggestions there are, the more good suggestions will be obtained. Moreover, one member's ideas will spark a related idea from a fellow member, and this in turn will spark an idea in another team member. To achieve this, participants are encouraged to generate as many suggestions as possible. Comments, particularly negative ones, serve to act as a damper during this process. For example, if a team member's ideas are shot down by others, this team member will tend to withdraw from contributing further ideas. It is advisable that no criticism of any suggestion by word, tone of voice, gesture or any other method of indicating rejection be allowed in order to keep the ideas flowing. All suggestions, no matter how apparently stupid, are recorded and none of them are rejected on the grounds of apparent irrelevance (SAVE, 1998a). Some of the basic rules of the brainstorming process are listed in Table 2.8.

Table 2.8 Brainstorming Process

	Basic Rules of the Brainstorming Process
1	The problem under study should be described to the team in advance. This description normally occurs broadly during function analysis in the information phase.
2	A positive environment should be established by the facilitator to embark on idea generation.
3	The group should be relatively small (e.g. up to eight members) and should consist of members from diverse backgrounds. This diversity is achieved by multi-disciplinary nature of the VM team. It may be also beneficial to include team members from both sexes because it is widely held that male/female thought processes differ in some ways. In brainstorming, any such difference is an advantage.
4	Quantity and not quality of ideas are encouraged, on the premise that the more ideas that are generated the greater the likelihood of valuable ideas.
5	Judgment of ideas is prohibited.
6	The combination and improvement of ideas is encouraged. To enable this process, the ideas are written on flip charts for all team members to see as they are generated.

Adapted from Norton and McElligott, 1995.

Evaluation Phase

In the evaluation phase, the ideas and concepts from the creativity phase are explored further and the most feasible ideas are developed into tangible value improvements (see Table 2.9). In order to highlight the best ideas for further study, the collected ideas are examined are rated in accordance with how well they meet the economic and non-economic evaluation factors determined in the pre-workshop stage (SAVE, 1998c; Norton and McElligott, 1995). The process typically involves several steps, which are summarised in Table 2.10.

Figure 2.3 shows the combined analysis matrix method, which combines criteria scoring matrix and alternative analysis matrix together. This method first determines the evaluation criteria according to their relative importance, and then scores

Table 2.9 Evaluation Phase

Typical Objective	Typical Questions	Typical Activities/ Techniques
Rank and Rate Alternative Ideas Select Ideas for Development	What do alternatives cost? What alternatives are functional? What ideas link together? Acceptability of options?	Rating/ weighting Life cycle costs Multi-disciplinary input Group/team interaction Common/corporate sense

Table 2.10 Key Steps of the Evaluation Phase

Key Steps of the Evaluation Phase
1   Eliminate nonsense or 'thought-provoker' ideas.
2   Group similar ideas by category within long term and short term implications. Examples of groupings are electrical, mechanical, structural, materials, special processes, etc.
3   Have one team member agree to 'champion' each idea during further discussions and evaluations. If no team member so volunteers, the idea or concept is dropped.
4   List the advantages and disadvantages of each idea.
5   Rank the ideas within each category according to the prioritised evaluation criteria using such techniques as indexing, numerical evaluation, and team consensus.
6   If competing combinations still exist, use matrix analysis to rank mutually exclusive ideas satisfying the same function.
7   Select ideas for development of value improvement.

Adapted from SAVE, 1998c.

the alternatives on the basis of the weighted criteria to determine those that are optimal. The key steps of this method are presented in the Table 2.11. According to SAVE (1998), if none of the final combinations satisfactorily meet the criteria, the value study team should return to the creative phase.

Table 2.11 Combined Matrix Analysis Method

Key Steps of the Combined Matrix Analysis Method		
1	Criteria Scoring Matrix	Select the criteria and list them on the left hand side of the criteria scoring matrix. There is theoretically no limit on the number of criteria used but due to time constraints it is rare to allocate more than twenty in a VM study. It is very important that each criterion is wholly independent of the others because any overlaps will skew or bias the results.
2		Assign a relative importance to each of the established criteria – to decide which criteria are more important compared with the others.
3		The scores for each letter may be added up to provide a relative weighting for each of the criteria.
4	Alternatives Analysis Matrix	Evaluate each alternative in turn against each of the criteria on a scale of 5: Excellent, 4: Very Good, 3: Good, 2: Fair, and 1: Poor.
5		Convert the scores of the alternatives against criteria to weighted scores, which will reflect the varying degrees of importance of each of the criteria.
6		Add up the weighted scores of each of the alternatives in the total column on the right side of the alternatives analysis matrix. The alternative with the highest score is then taken to be the optimal solution.

Weighted Evaluation

Project:

Date: ———

■ Architectural   ■ Structural   ■ Mechanical   ■ Electrical

VE No.: ———

Criteria

Criteria Scoring Matrix

How Important

4 – Major Preference
3 – Above Average Preference
2 – Average Preference
1 – Slight Preference
- Letter/Letter
- No Preference
  Each Scored One Point

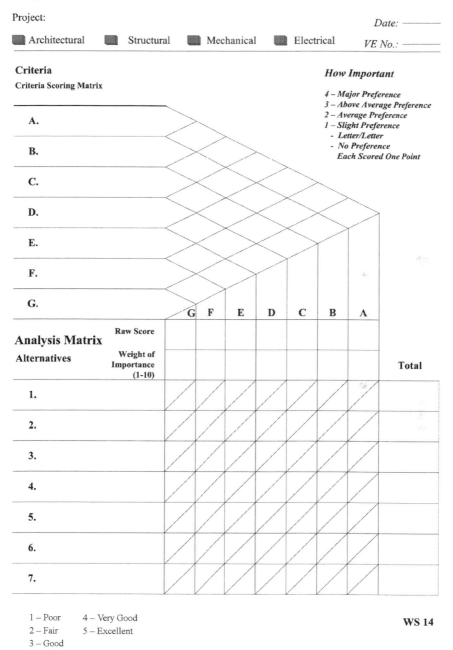

1 – Poor	4 – Very Good
2 – Fair	5 – Excellent
3 – Good	

WS 14

Figure 2.3 A Combined Matrix Analysis Method

Adapted from Norton and McElligott, 1995.

Development Phase

In the development phase, the objective is to select and further prepare the best alternative(s) for improving value, which were found earlier. The selected ideas are developed and turned into written recommendations for implementation (see Table 2.12). The champion of the given alternative prepares a package with the necessary data on technicality, cost and schedule for the designer and client to make their initial assessment with and decide on its feasibility and implementation (SAVE, 1998a). The next phase, the presentation phase, includes the key steps as shown in Table 2.13.

Analysis is usually conducted in both the analysis phase and development phase. Can you describe the differences between them?

Presentation Phase

The presentation phase is about obtaining agreement and commitment from the involved partners. The designers, project sponsors and other stakeholders give the go-ahead for the recommendations to proceed forward to implementation. All of the recommendations are summarised and put into a final proposal that is presented by the VM study team to all the decision makers for their approval. The decision makers then either give their approval or ask for additional information from the VM team. The final report prepared by the VM team documents the alternatives that have been proposed along with supporting data, and it confirms the implementation plan that was approved by management. Each final report is unique to the specific VM study and so the exact form of a report will differ (SAVE, 1998a).

Table 2.12 Development and Presentation Phases

Typical Objective	Typical Questions	Typical Activities/ Techniques
Conduct Benefit Analysis	What are the value	Cogent report and
Complete Technical	improvements?	executive summary
Data Package	Why change from status quo?	Clear action resolution
Create Implementation Plan	What further actions are	Co-ordinate actions
Prepare Final Proposals	needed?	Action plan and follow up
Present Oral Report	What decisions are required?	
Prepare Written Report		
Obtain Commitments for		
Implementation		

Table 2.13 Key Steps of the Presentation Phase

Key Steps of the Presentation Phase

1	Beginning with the highest ranked value alternatives, develop a benefit analysis and implementation requirements, including estimated initial costs, life cycle costs, and implementation costs taking into account risk and uncertainty.
2	Conduct performance benefit analysis.
3	Compile technical data package for each proposed alternative: • written descriptions of original design and proposed alternative(s) • sketches of original design and proposed alternative(s) • cost and performance data, clearly showing the differences between the original design and proposed alternative(s) • any technical back-up data such as information sources, calculations, and literature • schedule impact
4	Prepare an Implementation Plan, including proposed schedule of all implementation activities, team assignments and management requirements.
5	Complete recommendations including any unique conditions to the project under study such as emerging technology, political concerns, impact on other on-going projects, marketing plans, etc.

Adapted from Norton and McElligott, 1995.

2.6   Post-Workshop Stage

The objective of the post-workshop phase is to assure the proper implementation of the approved changes as recommended by the VM group. Assignments are carried out to track progress and collect feedback (SAVE, 1998a).

2.7   References

Akiyama, K. (1991). *Function analysis: systematic improvement of quality and perform-ance*. MA, USA: Productivity Press.

Commission of the European Communities (1995). *Value Management Handbook*. Luxembourg: Office for Official Publications of the European Communities.

Crow, K.A. (2002). *Value analysis and function analysis system technique*, DRM Associates. www.npd-solutions.com/va.html

Dell'Isola, A.J. (1982). *Value engineering in the construction industry*, 3rd edn. New York: Van Nostrand Reinhold Company Inc.

Dell'Isola, A.J. (1997). *Value Engineering: Practical Applications... for Design, Construct-ion, Maintenance and Operations*. Kingston, MA: RSMeans.

Masson, J. (2001). *The use of functional performance specification to define information systems requirements, create RFQ to suppliers and evaluate the supplier's response*. Conference paper of the SAVE Annual Congress, May.

McGeorge, D. and Palmer, A. (1997). *Construction Management: New Directions*. Oxford: Blackwell Science.

Norton, B.R. and McElligott, W.C. (1995). *Value Management in Construction: A Practical Guide*. Basingstoke: Macmillan Press.

SAVE International (1998a). *Function: definition and analysis*. www.value-eng.org/ pdf_docs/monographs/funcmono.pdf

SAVE International (1998b). *Function analysis systems technique: the basics*. www.value-eng.org/pdf_docs/monographs/FAbasics.pdf

SAVE International (1998c). *Value methodology standard*, 2nd edn. Northbrook: SAVE International.

HKIVM (2004). The Hong Kong Institute of Value Management, *The Value Manager*, 10(4).

3   Group Dynamics in VM

Geoffrey Q.P. Shen, Ann T.W. Yu, and Jacky K.H. Chung

3.1 Introduction

This chapter provides a facilitation training resource to enable students and VM group members to understand and assume the role of group facilitator. It also explores how group effectiveness can be vastly increased through synergy.

3.2 Learning Objectives

Upon completion of this chapter, you should be able to:

1. Recognise and understand the dynamics generated through the group process
2. Establish and promote group dynamics that satisfy members' socio-emotional needs and help groups achieve goals
3. Explain how a team operates and how it can become more successful

3.3 Teamwork

Group dynamics is a large body of research in psychology and the social sciences. A thorough understanding of group dynamics is useful for working effectively within any type of group. Although many theories describe the function of groups, fundamental to all these theories is an understanding of groups as social systems.

Individuals in a group will benefit from understanding the basic principles of what happens in groups of people, especially when they must work closely together. This area of study was labelled 'group dynamics' by sociologists who attempted to discover what typically goes on in groups (Rees, 1997).

Definition of Group

> **What is your own definition of a group and how many types of group are there?**

A group can be defined in psychological terms as any number of people who interact with one another; are psychologically aware of one another; and perceive themselves to be a group

(Schein, 1980)

Types of Groups

Groups can be divided into two categories: formal and informal

(Schein, 1980)

Types of formal Groups

Managers create formal work groups to perform organizational tasks. There are two types of formal group.

- Permanent formal groups – are unlikely to be disbanded although individual group membership may change over time. An example would be a standing committee.
- Temporary formal groups – consist of project teams and task forces. They can exist for a long period of time, but members know that at some time in the future they will cease to exist as a group. They have a life-cycle formation through to cessation. An example would be a value management team.

Types of Informal Groups

Informal groups exist for social reasons rather than organizational task related reasons. They are often related to physical proximity, which increases the likelihood of social interactions.

- Horizontal cliques – group members are near equal organizational status level.
- Vertical cliques – group members may be from different organizational status level within the same department or team.
- Mixed or random cliques – group members are from different organizational status level, department and locations.

Group Composition

> **What qualities does an effective group need? Make a list!**

A group performance depends on the qualities of the individuals who are performing the task. The optimum qualities for a group to possess are as follows:

- A combination of knowledge, skills and abilities that match task requirements

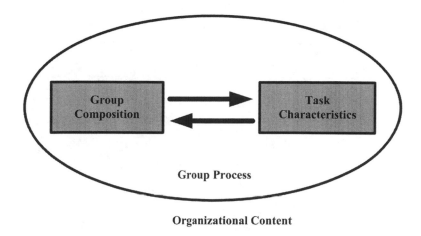

Figure 3.1 Model of Team Interaction

Adapted from Daniel Levi, 2001.

- Members with the authority to represent the relevant parts of the organization and the power to implement the group's decisions
- The necessary group process skills to operate effectively

Group Dynamics

To understand the social structure of groups and develop effective group work skills, four areas of group dynamics are helpful (Toseland and Rivas, 2001), these are:

- Communication
- Cohesion
- Social control mechanisms, i.e. norms, roles, and status
- Group culture

Communication and Interaction

Social interaction is a term for the dynamic interplay of forces in which contact between persons results in a modification of the behaviour and attitudes of the participants

(Northen, 1995)

Communication contains (1) encoding of a person's perception, thoughts, and feelings into language and other symbols, (2) transmission of these symbols or language and (3) decoding of the transmission by another person

(Northen, 1995)

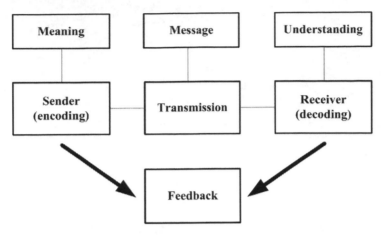

Figure 3.2 A Model of Communication

Adapted from Northen (1995).

Effective communication and intervention enables the group to understand the 'bigger picture' and achieve desired goals.

Group Cohesion

> Group cohesion is the interpersonal bonds that hold a group together
> (Hogg, 1992)

> The joining together to work as a team creates a sense of cohesiveness
> (Guzzo and Dickson, 1996)

According to Cartwright (1968), four interaction variables determine members' attraction to a particular group:

- Need for affiliation, recognition, and security
- Group participation, resources and prestige available
- Beneficial and detrimental results of expectations of group members
- Expectations of non-group members which may affect the group

Cohesion in a group comes from the need for affiliation felt by the members. If a member feels that their relationships outside of the group are unsatisfactory or non-existent, he or she may feel the need to socialise more.

In a cohesive group, the members' accomplishments are recognised and a sense of competence is promoted among the members.

A group's cohesion can be accounted for by looking at the incentives that are a result of the membership of the group. Reasons for joining a group may be the expectation of meeting and getting to know certain people.

Social Control

Social control refers to the process in which all members of a group get sufficient compliance and conformity among each other to enable it to function in an orderly manner. Social control is affected by such factors as the norms, roles and status of the individual group members.

The amount of social control needed differs from group to group. In groups where strong social control is required for the group to function effectively, the members of the group may have to relinquish a great deal of their own freedom and individuality.

GROUP SIZE

Hunt (1986) stated that for optimum activity, group size should be between six and ten people. Otherwise, with larger groups, there is a danger that the group will fragment and small cliques will form.

GROUP NORMS

These are codes or rules of behaviour (Hunt, 1986) that are generated from group interactions and the shared expectations of contributors. The longer the group has been operating, the more likely strong norms will develop.

GROUP ROLES

Each person has expectations about how they personally will behave in a group and also expectations and assumptions about other people's behaviour (Hunt, 1986).

According to Belbin's Team Role (Belbin, 1981), individuals contribute to a team in two ways:

- Functional role – contributing to the task, through specific technical or professional expertise
- Team role – contributing to the process, through behaviour, which facilitates progress

Belbin's Team Role contains eight dimensions (Belbin, 1981):

- Plant
- Resource investigator
- Shaper
- Implementer
- Teamwork
- Monitor-Evaluator
- Completer-Finisher
- Chairman

By understanding the team roles, team members can:

- understand their own contribution
- understand other members' contributions
- agree how to split tasks to minimize clashes between individuals competing for the same roles
- identify areas where attention is needed

GROUP STATUS

The status refers to an individual member's position within a group, an evaluation and ranking done relative to the other members of the group. The prestige of a member can also determine the person's status in the group. The status of a person can influence the behaviour in a group, for example, members with a low status level are less likely to conform to the group norms, because they have less to lose by deviating from expected behaviour. High status members however, perform many valuable services and usually conform to valued group norms when they are establishing their position in the group (Nixon, 1979).

Group Culture

Group culture refers to values, beliefs, customs and traditions shared between group members (Olmsted, 1959). Each individual member still has their own set of values that come from their past experiences and from their ethnic, cultural and racial heritage. Group cultures are established more rapidly in groups with a homogenous membership base. When the group members share common past life experiences and a similar set of personal values, their personal perspectives merge faster to form a coherent group culture.

3.4   Team Development

> **What is the team development process and how many stages does it include?**

According to Northen (1995), a stage is a differentiable period or a discernible degree in the process of growth and development.

The process of team development is usually summarized by five stages: forming, storming, norming, performing, and adjourning.

Forming

The formation stage of team development happens at the very beginning when a group of individuals first come together and form a team and start to think about themselves as members of a team.

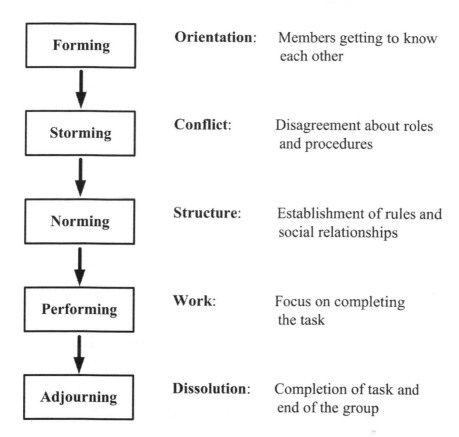

Figure 3.3 Stage of Group Development Forming
Adapted from Daniel Levi, 2001.

Storming

The second stage is storming, which involves the emergence of conflict between team members as they begin to disagree with one another. Conflict during storming can be beneficial or detrimental. A moderate amount of conflict is needed to bring out the process of compromise.

Norming

Third phase, norming, is a development phase. The team members have joined together and established a sense of belonging in the group. Cohesion within the group comes as a result of this development stage, and a group norm is developed. The norm helps regulate the behaviour of the members. Norming increases the ability of the team to stay together and focus on accomplishing their goals.

Performing

The fourth phase, performing, comes about when members start to work towards accomplishing their collective objective, through drawing on internal synergies between the members. Social relationships develop and team members increase their understanding of each other. This leads to a strong sense of commitment to the team, which ensures that the team strives towards high performance.

Adjourning

The last phase of the team development occurs, when the team starts to dismantle and the individual members move on to other activities, sometimes unexpectedly. External factors can lead to a team dissolving and terminating before they have met their goals and objectives. Such factors can be mergers and downsizing on a corporate level.

3.5   Effective Teamwork

To effectively function, a team must above all focus on collaboration and trust, and must commit to a democratic practice. When team members trust each other, the inevitable conflict becomes an opportunity to foster new thinking and creative ideas. Without trust, conflict can disable a team's progress toward its goals. A Value Management team coordinator's function is to unlock and guide the team's problem solving efforts toward the project.

Obstacles to Effective Teamwork and Effective Problem Solving

> **What are the key obstacles to effective teamwork? Please compare these obstacles with the blocks to creativity described below.**

Perceptual Obstacles (Adams, 1987)

- cannot see the problem clearly
- cannot see the information needed to solve the problem

Emotional Obstacles

- basic human psychological functioning of needs and desires

Cultural and Environmental Obstacles

- obstacles caused by social conditioning, such as shyness in public speaking and fear of looking foolish
- immediate social and physical environment

Intellectual or Expressive Obstacles

- difficulty in expressing ideas or concepts due to language problem or lack of specific terminology etc.
- insufficient or incorrect information

Group Problem Solving Techniques

According to Adams (1987), problem-solving techniques demonstrate two qualities:

- fluency – the number of concepts or ideas that can be generated in a period of time; and
- flexibility – the diversity of ideas generated.

The most commonly used techniques in VM are brainstorming, synectics, and Edward de Bono's PMI, CAF, and TEC techniques.

Brainstorming (Adam, 1987; Be Bono 1970)

- most common technique used for problem solving
- especially best for simple problems or problems that are well-defined

Synectics (Adam, 1987)

- more complex than brainstorming
- requires a high level of technical expertise
- especially applicable to value management project

PMI Technique (De Bono, 1982)

- P – plus or goal point
- M – minus or bad point
- I – interesting point
- P, M, I are effective for scanning and objective thinking

CAF technique (De Bono, 1970)

- Consider All Factors
- Especially effective for system oriented perspective on a project

TEC (De Bono, 1970)

- T = Target, Task of thinking: identify the focus and function
- E = Expand, Explore: expand the problem and enrich functional definition
- C = Content, Conclude: draw conclusions and identify outcomes

Negotiation

> **What is the definition of negotiation? What are the common types of negotiation skills that you use?**

Negotiation is a social process involving compromise through persuasion to resolve self-interest and conflict issues and reach consensus. It takes place between:

- individuals
- an individual and a group
- groups

Types of Negotiation

- Resolving grievances
- Counselling
- Bargaining
- Problem solving

Negotiation Skills (Fisher and Urry, 1987)

PRINCIPLES OF NEGOTIATION

- separating people from problems
- targeting interest rather than positions
- generating copious options before deciding
- judging results against objective criteria

ROLE REVERSAL

- adopt the perspective of the 'other side'
- difficult to apply in practice since it requires detachment and objectivity

COMMON GROUND

- agreement area between parities
- the broader the area, the better agreement is reached

OBJECTIVE CRITERIA

- any negotiation should be judged against objective criteria (e.g. moral standards, market prices, traditions, etc.)
- parties to an agreement can judge its merits, along with the vested interests that are present in the negotiation situation

NEGOTIATION TEAM

Kennedy *et al.* (1984) suggested that individuals comprising a negotiation team be allocated specific tasks:

- leader – conduct or present the problem, championing a side
- summarizer – clarification
- observer – listen and watch

BEST ALTERNATIVE

Fisher and Urry (1987) suggested the use of BATNA (Best Alternative To a Negotiated Agreement) for:

- working out a best alternative
- developing a list of future action if no agreement is reached

3.6 References

Adams, J.L. (1987). *Conceptual blockbusting: a guide to better ideas*, 3rd edn. Harmondsworth, UK: Penguin.

Belbin, R.M. (1981). *Team roles at work.* Oxford, UK: Butterworth Heinemann.

Cartwright, D. (1968). The nature of group cohesiveness, in D. Cartwright and Zander (eds), *Group dynamics: research and theory*, 3rd edn, pp. 91–109. New York: Harper & Row.

De Bono, E. (1970). *Lateral Thinking.* Harmondsworth, UK: Penguin.

De Bono, E. (1982). *de Bono's thinking course.* London: Holland.

Fisher, R. and Urry, W. (1987). *Getting to yes: negotiating an agreement without giving in.* London: Arrow.

Guzzo, R.A. and Dickson, M.W. (1996). Teams in organizations: recent research on performance and effectiveness. *Annual Review of Psychology*, 47: 307–338.

Hogg, M. (1992). *Culture's consequences: International differences in work related value.* Beverly Hills, CA: Sage.

Hunt J.W. (1986). *Managing People at Work*, 2nd edn. London: McGraw-Hill.

Kennedy, G., Benson, J. and McMillan, J. (1984). *Managing Negotiations*, 2nd edn. London: Hutchinson.

Levi, D. (2001). *Group Dynamics for Teams.* Sage Publications.

Nixon, H. (1979). *The Small Group.* Englewood Cliff, NJ: Prentice-Hall.

Northen, H. (1995). *Clinical social work knowledge and skills*, 2nd edn. New York: Columbia University Press.

Olmsted, M.S. (1959). *The Small Group.* New York: Random House.

Rees, F. (1997). *Teamwork from start to finish: 10 steps to results.* San Francisco, CA: Pfeiffer.

Schein, E.H. (1980). *Organizational Psychology*, 3rd edn. Englewood Cliff, NJ: Prentice-Hall.

Toseland, R.W. and Rivas R.F. (2001). *An introduction to group work practice.* Allyn & Bacon.

4 Group Facilitation in VM

Geoffrey Q.P. Shen, Ann T.W. Yu,
and Jacky K.H. Chung

4.1 Introduction

This chapter explains the concepts of facilitation and creativity, and describes how these elements can be properly implemented in VM studies.

4.2 Learning Objectives

Upon completion of this chapter, you should be able to:

1. Define the concept of facilitation
2. Facilitate a group and explore how to vastly increase group effectiveness through synergy
3. Define the concept of creativity
4. Manage the skills of encouraging creativity in groups
5. Identify the key obstacles to creativity and the methods used to overcome them

4.3 Introduction to Facilitation

> **Are you familiar with the process of facilitation? If so, how would you define it? Write a definition before continuing.**

Facilitation is about process, 'How you do something', rather than the content, 'What you do'

(Hunter *et al.*, 1996)

To facilitate is equal to make easy or more convenient

(Hunter *et al*, 1996)

The facilitator is a guide through the processes. Someone who helps make the process easier and more convenient. The facilitation process is about movement, i.e. moving something or progressing from A to B. The role of the facilitator is to guide the team or group towards their destination and make getting there easier.

We can facilitate another person, a group or ourselves. Before facilitating a group, we have to facilitate ourselves in terms of our own internal and external processes.

A powerful facilitator needs to train in self-facilitation, facilitation of others, and facilitation of a group.

The triangle shown in Figure 4.1 illustrates the relationship between self-facilitation, facilitating another, and facilitating a group in meeting its stated purpose. The centre of the triangle is the agreement on how the group will work together, including the culture, internal environment, and group contract. The external environment is all the influences, formal and informal, which affect us as citizens and members of organizations and cultures. These influences will come into the group subconsciously and may lead to misunderstandings and conflict. Although misunderstandings cannot be avoided entirely, the best way to minimize them is to create an explicit written agreement as to how members of the group will work together (Hunter *et al.*, 1999).

Facilitating Yourself

Facilitation aims to maximise the potential of individuals and thereby to maximise the efficiency and effectiveness of a group of people. It is about self-awareness. The way we grow and develop as conscious human beings is by facilitating ourselves and being facilitated by others. This happens on an emotional, mental and spiritual level.

**FACILITATION GROUP TO ACHIEVE
ITS PURPOSE**

EXTERNAL
ENVIRONMENT

EXTERNAL
ENVIRONMENT

CULTURE

INTERNAL

NVIRONMENT

CONTRACT

FACILITATE SELF

FACILITATE ANOTHER

Figure 4.1 A Model of Facilitation

Adapted from Hunter *et al.*, 1999.

Much of our physical development is also self-facilitated. Our body shape, tone and stamina are affected by exercise. We can modify our body to build up certain muscle groups for the performance of special functions, as sportspeople, musicians, dancers and singers do (Heron, 1993).

Being with Yourself

The first step in any facilitation is to be comfortable and at home with yourself, your own space, thoughts and feelings. Before we get involved in facilitating other people we need to accept ourselves and be self-aware. Facilitating ourselves is about growing, developing and training ourselves but not about drastically altering ourselves. Accepting oneself is the biggest difficulty for most people.

Empowering Ourselves

We can empower ourselves through self-facilitation. Empowerment is about recognizing when we are 'in our power' and how we experience this and bring it into all aspects of our life.

Facilitating Others

Being comfortable and at home with people is an important step in facilitating others, which is essential for facilitating groups. Facilitating other people is easier than facilitating oneself, as we can see other people's patterns or blocks more easily than our own.

Facilitating a Group

A group is an entity rather than a collection of individuals. It is a living system with its own personality, potential and limitations. Group members are not just joined physically but they are joined emotionally, intuitively, intellectually and spiritually.

Some people find being in a group to be a challenge, as they fear losing some of their own identity and autonomy. Another fear associated with group work is the loss of freedom or a fear of being dominated by others.

Group members might occasionally be under powerful peer pressure to comply with specific group values and internal behaviour. A group focused on cooperative behaviour might be particularly susceptible to this, as there is pressure on everybody to reach agreement on major decisions.

A facilitator needs to be aware of:

- The individual member's need for freedom
- The collective need for co-operation in the group
- The group as a closed system or organization, where a particular culture and personality dominates

A facilitator should therefore be comfortable with others in a group: be comfortable with the way the group works, its culture and personality, and know the value of the group as a whole.

To facilitate a group takes a certain degree of fearlessness. It takes sufficient self-awareness to realize uncertainty and a willingness to go with the group flow. A facilitator needs commitment to the group fulfilling its purpose, to tap into the group's collective mind and creativity (Hunter *et al.*, 1996).

Facilitator's Role in the Group

The facilitator should remain independent and thus only be responsible and accountable to the group. The group in turn gives the facilitator their trust and a contract under which he is to guide the group towards fulfilling their purpose (Hunter *et al.*, 1996). The facilitator should be familiar with the expectations of the group towards the facilitator's role, and should share his or her own expectations, values and ways of working with the group. This can help mutual understanding and create an open atmosphere of sharing within the group.

Facilitation Guidelines

- Group facilitation is the art of guiding the group process towards the agreed objectives, so a facilitator only guides the process rather than being involved in the content. They can intervene to protect the group process and keep the group on track to complete the task, but only when necessary.
- A group is more than the sum of its individual parts; a group builds on the synergies that can be achieved within it. An effective facilitator must work around individual members past experiences and give them the right encouragement so they can achieve great results.
- The facilitator must to tap into the group synergy and exploit for positive gains. The facilitator has to trust the group and the resources they have available and work through any process issues towards using these resources to accomplish their tasks.
- Facilitation is about appreciating each group member and encouraging their full participation in having the group achieve their tasks effectively and efficiently.
- The facilitator should ensure that the physical space is safe and guarded from interruptions and intrusion.
- The facilitator should always remember the purpose of the group. It can be useful to have this displayed visibly and publicly at each meeting.
- It is important for the facilitator to be adaptable.
- The facilitator should keep intervention to a minimum and should only intervene in a group discussion when:
    1. Progress towards fulfilling the task is impeded or group synergy undermined in some way.
    2. Discussion is becoming irrelevant or a group member is blocking progress due to past preconceptions or inappropriate ideas.

3.   A situation becomes physically dangerous.

- The facilitator should always seek consensus in a group using a collective decision-making processes, unless there is a group agreement to do otherwise.

Beside the facilitation guidelines given above, can you suggest one or two guidelines in accordance with your own working experience?

Synergy

Synergy is about:

> tapping into group energy so that the group members are able to accomplish more than they thought possible. Tapping into group energy increases dramatically the speed at which a group takes action. Synergy is about flowing and working together harmoniously, and about co-ordinating action and being inspired by one another. Increasing group and facilitation skills will open up the pathway to experiencing synergy in your group. Group skills increase the level of co-operation in a group. Members begin to listen to one another in a new way, and start to recognize when they are in tune with others
>
> (Hunter *et al.*, 1996)

4.4   Introduction to Creativity

Definitions of Creativity

What is creativity? Try to define it yourself in a sentence or two before reading on.

> Creativity is a distinguishing characteristic of human excellence in every area of behaviour
>
> (Torrance, 1979)

> Creativity is the application of imaginative thought, which results in innovative solutions to many problems. These solutions usually emerge either through the synthesis of a large amount of material or by searching for and finding connections and patterns that others fail to see
>
> (Goodman, 1995)

Characteristics of Actively Creative People

Leading qualities that are identified in the literature (e.g. Garnham and Oakhill, 1994; Hellriegel and Slocum, 1975; Kreitner, 1980) include:

- An ability to find the right problem to address
- An ability to defer judgment on possible solutions
- A desire for originality
- A tendency to resist social forces to conform
- A tolerance of ambiguity
- An ability to adopt a positive attitude
- A strong belief in personal creativity

Actively creative people have a talent for getting to the heart of a problem. They are not confused by detail and by the need to invoke standard approaches. They enjoy experimenting with ideas and, whilst the answer sometimes comes to them in a flash, they are also prepared to toy with a challenge for long periods of time. They are keen to discover new solutions and willing to innovate even if this means having to resist social paradigms and mind-sets (Sternbergs, 1988).

> **According to the characteristics described above, do you think that you are a creative person? If not, how could you develop your own creativity? Thinking about questions like this also develops your awareness.**

Encouraging Creativity

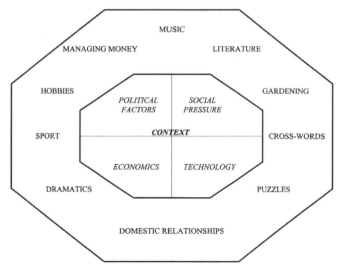

Figure 4.2 Activities that entail Personal Creativity
Adapted from Goodman, 1995.

Contextual Factors

Anyone is capable of responding creatively. Figure 4.2 illustrates some of the ways that ordinary people exhibit creativity. The degree to which personal creativity occurs depends on a complex set of environmental or contextual factors, such as social pressures and technology. Political factors can also influence creative responses, such as thinking of novel ways to get around existing or proposed legislation (Goodman, 1995).

Organization and Creativity

Organizations cannot easily respond creatively, but they can actively encourage their individual employees to do so. Thus the incidence of creative activity is contextually governed by the degree of group and/or organizational support; the 'vibes in the air' have to be convincing. When this is the case, people will invoke the processes and creative responses will occur (Goodman, 1995).

Managers and Creativity

Managers have to make a real commitment to openly welcome creative responses and then champion them. They need to adopt a new approach and develop individual personal relationships with their staff (Goodman, 1995).

Capturing Creativity

To think creatively, individuals have to really believe that they enjoy the freedom to think in ways different from the accepted norms. For this to happen regularly, individuals should be given a succession of positive challenges and be constantly encouraged to perform to the best of their abilities. To achieve a consistent market edge, creative thinking must be released and continuously supported (Goodman, 1995).

CREATIVE GROUPS

Helping Others to Experience Creativity

Whilst native curiosity, persistence and a willingness to persevere will bring its reward at the individual level, it can nonetheless be difficult for individuals to share their discovery and experiences with others. This is a result of a complex set of factors including a degree of self-consciousness, not wanting to look silly to colleagues, and the personal approaches that all of us have to make in order to find and appreciate the creative force (Goman, 1989).

Thus one individual cannot lead another by reason alone to a realization of the potential of creative force. However, if a curious individual can temporarily suspend his or her personal prejudices and adopt an open, inquiring regard, then a successful outcome will result. Creativity is a natural force that can convince doubters at a deeply personal level once they experience it. Sharing creativity cannot be achieved through words alone and it is a disservice to the cause for individuals to attempt this. The force exists; the issue is how to assist others to find it for themselves.

Establishing a Creative Climate

The first task of a creative facilitator is to stimulate individuals to become curious about creative thinking. The creative force is omnipresent but not always obviously active in the behaviour of individuals. Just as individuals can choose their behaviour, they can choose to harness their natural creativity. Thus it is a matter of individual choice, and this choice is heavily dependent on dominant contextual factors, such as environment ambience, which a creative facilitator needs to give careful attention to (Ekvall, 1983). Creative facilitators should also lead by example. They have to:

- Generate a supportive atmosphere in the established physical setting by purposefully signalling to work groups that they really mean business.
- Encourage small group formation (e.g. seven) and balance the group members in terms of functional and general skill and length of service.
- Encourage all group members to contribute, affirming the philosophy 'all for one and one for all'. Discourage any tendency for an individual to assert too great an influence on the group (Goodman, 1995).

CREATIVE PROBLEM SOLVING (CPS)

Problem Evaluation

Problem evaluation is an initial inquiry into the nature of a problem requiring a creative management response (CMR). Many managers spend considerable time trying to identify the real problems. Operational intuition often tells them that something is wrong or needs attending to, but not always what needs to be solved.

Idea Generation and Development

Managers need to find the best method of solving their problems under their individual circumstances, especially when they are concerned with competition, cost or quality. VanGundy (1988) proposed that the selection of a CPS tool should match the characteristics of the problem in terms of complexity and importance.

Realization Stage

The realization stage includes all the actions necessary to implement a selected creative management response (Kolb, 1984; Kreitner, 1980).

Figure 4.3 summarizes the main factors that influence the expression of the creative management response in group situations.

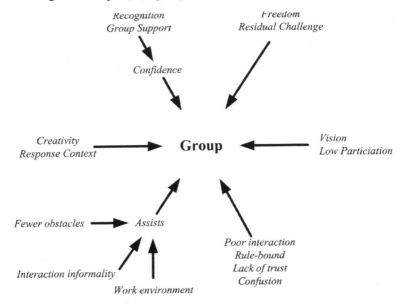

Figure 4.3 Main Factors Influencing the Expression of the Creative Management Response
Adapted from Goodman, 1995.

Obstacles to Creativity

Obstacles to Personal Creativity

Obstacles to personal creativity can be for the reasons summarized in Table 4.1.

Table 4.1 Obstacles to Personal Creativity

Obstacle	Description
Tiredness	Reluctance to put in the necessary time to learn a new skill.
Anxiety	Often transferred after working hours to home life. Many people find it difficult to relax away from work and free their minds to attempt anything new.
Negativity	Frustration with management performance or a reaction to a pronounced negative culture within the work environment.
Fear of failure	Frequently felt by individuals working in both large and small organizations. Many do not realize that it is possible to practice creativity at home in a safe environment.

Adapted from Goodman, 1995.

Obstacles to Group Creativity

Table 4.2 shows the obstacles to group creativity along with some suggestions for overcoming them.

Table 4.2 Obstacles to Group Creativity

Obstacle	Suggestion
Negative attitudes	Quietly get on and try some CPS tools
Politics	Stand your ground, argue to judge result
Fear of exposing poor teamwork of other groups in the organization	Operate tactfully, tackle problems, resist negativity or scepticism
Myopia (the short sightedness of seeing only the short term problem)	Explain contextual factors affecting the problem to all group participants
Concern over choosing the right problem to obtain recognized success	Start simple, select progressive issues.
Low creativity	Instructions and guidelines from a creative facilitator
Lack of trust	Do your best to earn it.
Poor problem-solving ability	Introduce some basic tools (e.g. brainstorming)
Unclear aims	Try to clarify
Reward structure viewed as unfair	Show your appreciation and seek to influence the system

Adapted from Goodman, 1995.

4.5 References

Ekvall, G. (1983). *Climate, structure and innovativeness of organization: a theoretical framework and an experiment.* Stockholm, Sweden: FA rådet, The Sweden Council for Management and Organizational Behaviour.

Garnham, A. and Oakhill, J. (1994). *Thinking and Reasoning.* Oxford: Blackwell.

Goman, C.K. (1989). *Creative Thinking in Business.* London: Kogan Page Ltd.

Goodman, M. (1995). *Creative Management.* London: Prentice Hall.

Hellriegel, D. and Slocum, J.W. (1975). *Managerial problem solving styles.* Columbus, OH: Business Horizons.

Heron, J. (1993) *Group Facilitation: Theories and Models for Practice.* London: Kogan Page Ltd.

Hunter, D., Bailey, A., and Taylor, B. (1996). *The Facilitation of Groups.* Aldershot, UK: Gower Publishing Limited.

Hunter, D., Bailey, A., and Taylor, B. (1999). *Handling Groups in Action.* Aldershot, UK: Gower Publishing Limited.

Kolb, D.A. (1984). *Experiential Learning.* Englewood Cliffs, NJ: Prentice Hall.

Kreitner, R. (1980). *Management: A Problem-Solving Process.* Boston, MA: Houghton Mifflin.

Sternberg, R.J. (1988). *The Nature of Creativity: contemporary psychological perspective.* Cambridge: Cambridge University Press.

Torrance, E.P. (1988). The nature of creativity as manifest in its testing, in R.J. Sternberg (ed.), *The Nature of Creativity: contemporary psychological perspectives.* Cambridge: Cambridge University Press.

Van Gundy, A.G. (1984). *How to establish a creative climate in the work group.* Management Review, 24–38, August.

5 VM Implementation

Ann T.W. Yu, Geoffrey Q.P. Shen,
and Jacky K.H. Chung

5.1 Introduction

This chapter introduces the benefits and critical success factors of VM applications and explores the most frequently encountered difficulties in VM studies.

5.2 Learning Objectives

Upon completion of this chapter, you should be able to:

1. Demonstrate an awareness of the benefits and critical success factors of VM
2. Explain the difficulties in implementing VM studies in the construction industry and identify their causes

5.3 Benefits

The benefits of VM, in terms of better value and improved return on investment, can be achieved through the following.

Project cost savings

A well-administered VM programme should yield cost reductions in the order of 10–25 percent of total construction project cost where the additional cost of VM is estimated at less than 1 percent of the total construction project cost. VM literally pays for itself by elimination of unnecessary cost.

Timesaving in design, construction, approvals

Early application of VM will save design time by clarifying the scope of work, reducing false starts, and helping to prevent redesign and reconstruction.

Improved project management structures and systems

VM improves the project management structures and systems by bringing all the consultants together in a workshop so that they can discuss the project face-to-face. This is much better than the traditional system where the project manager communicates individually with the design consultants and the design consultants rarely talk to each other.

Consideration of options

The early application of VM will recognise the strengths, weaknesses, opportunities and threats created by the 'build' or 'no build' options. Any project should only be initiated after a careful analysis of need. Failure to do this will cause problems in the design and construction stages and, more importantly, it will cause problems in the operation and usage in the longer term.

Expedite decisions

There are number of key points in every construction project at which the client must make important decisions. When VM is applied at these stages, it helps to ensure that these decisions are taken in a way that is rational, explicit, accountable and auditable. This means that key project stakeholders are able to participate more fully and effectively in decision making, which increases their confidence in the process. It also means that they are more committed to a successful project outcome.

Minimising wastage

Clients must be aware of the need to minimise wastage in their construction projects. By careful consideration of different options for achieving the agreed objectives, VM may eliminate duplicating effort and abortive work in both design and construction.

Forecasting risks

Risk management is an integral part of VM. Even though it might seem irrelevant to identify and manage risk based on agreements that have not yet been made, it is however a positive thing as a strategic diagnosis of the risks can help set the objectives. Outline design proposals and project feasibility studies should feature in considerations of project risks.

Concentrating expenditure on adding value

VM works by making explicit objectives and value for money criteria for stakeholders. This helps to fully investigate the need for a project before making a financial commitment and provides a structured framework within which subsequent decisions can be taken in accordance with the objectives and criteria. It means that the project design is developed and continuously evaluated against need, so that value for money is achieved.

Improving communication and understanding

VM drives the team members beyond their habitual patterns and procedures. The basic physical limitations of an office set-up prohibits the free flow of information between various groups, whilst a VM programme provides a way of mobilising and uniting the individual talents of each member to achieve the project's designated objectives.

> **Besides the general benefits of VM described above, what are the specific benefits of VM application in the design phase?**

Benefits of VM Application in the Design Phase

Traditionally, the VM technique is used at the detailed design stage, when most of the essential design decisions have been made. Because of its successful performance when applied to building design, this technique was developed and recommended for use throughout the design of a building project. In theory, VM can be applied at any stage of the building design process. In practice however, in order to get maximum return on VM input, VM is usually applied to a project at the conceptual design and sketch design stages, and is rarely used in the detail design stage.

1. VM is a straightforward and effective approach

VM's organised approach is an incentive process that seeks out information, stimulates the thought process in a single session, and evaluates these thoughts and ideas to produce recommendations for implementation. Furthermore, the group creativity technique avoids criticism by separating the creative aspects of the study from the judgmental aspects.

2. VM programmes identify and remove unnecessary costs

Each design is in itself a creative thought process and there are an infinite number of combinations of designs, materials and methods that can be used to achieve the ultimate goal of a project. The real skill is finding the right balance between the cost, the performance, and the reliability of the project.

Each designer working on a project carefully develops a number of alternatives that can be used to perform a function. However, a designer may use his/her experience to by-pass many comparisons in the interest of meeting due dates and design budgets. Analysis of historical projects has shown that the depth and thoroughness of the design partly depends on the time and resources allocated (Zimmerman and Hart, 1982).

Given the time and resources, under proper project management, comparisons might be made to optimise the design. Additional information towards the project and an objective opinion by someone not involved in the design may change the outcome of previous comparisons.

3. VM facilitates effective communications and co-operations

VM drives the team members beyond their habitual patterns and procedures. A multi-disciplinary team motivates its members by the fact that they are participating in a group-thinking process, and working towards the same objective. Although a successful project design is dependent upon co-operation between various design groups (e.g. electrical, mechanical, architectural, structural), the physical limitations of an office set-up prohibits the free flow of information between them. A VM programme overcomes this restriction by providing a way of mobilising and uniting individual talents from each group to achieve the project's objectives. As Jones (1970) stated:

> the secret to the success of design is to organise the process so that each person is acting as much as possible adaptively and creatively, and so that we minimise the need to work to rule, shutting the mind to the evident effects of what one is doing
>
> (Jones, 1970)

4. VM works in a circumstance where comparisons are easy to make

A design is by nature a creative process: as Jones (1970) stated, 'design is to initiate changes in man-made things'. This process of creation is made in a step-by-step fashion. As a design develops, those pieces are pulled together into a whole. A deductive reasoning process is used, in which the major mental process is the recall of past knowledge. In contrast, when a VM programme is carried out, those pieces have already been pulled together and the inductive reasoning process is used to analyse the project. In this case, the major process is the comparison of alternatives, and it is easier to evaluate a given idea than to come up with the original concept (Zimmerman and Hart, 1982).

5. VM promotes progressive changes

VM tends to establish an atmosphere where changes can be made gracefully without anyone losing face. Many times self-interest allows our thoughts to be dominated by our desires, which is a regular occurrence in engineering practice.

6. The VM programme and the design have an identical goal

The identical goal is to provide a satisfactory design for the project, meeting the owner's requirements and providing the best balance between cost, performance and reliability.

Keeping this in mind, it is easy for the project engineer to realise that the VM team is a supplement to his design effort. A second look is taken at the design in an attempt to improve the cost impact of the project, rather than to criticise the designer's work

(Zimmerman and Hart, 1982)

Although this statement is probably true when VM is applied at the conceptual design stage, later applications of VM to a building design could lead to abortive design work and is unlikely to be accepted by the designers.

7. VM reinforces the value of a project

Although this comes last on the list, it is by no means least in importance. VM consists of a range of well-developed techniques such as function analysis, life cycle costing analysis, decision analysis, individual and group creative thinking, which reinforce the value of a project by enhancing its performance and/or reducing the overall cost. Although a VM programme itself costs money, a benefit of ten for every one dollar investment can be expected from a formal VM programme (Dell'Isola, 1982). The earlier in the design process VM can be applied, the bigger the savings that can be expected.

Limitations of VM Application in the Design Phase

The major criticism of VM is the time delays and the extension of the design programme that it causes. Some designers blame the VM exercise for adding to the time required for design. Although these critiques are not absolutely right, they do reflect limitations of the VM programme as summarised in Table 5.1.

Table 5.1 Limitations of VM Application in the Design Phase

No.	Limitations
1	If the VM workshop is not properly organised, the study can be fruitless. Some designers have complained that many suggestions in the VM proposal had previously been considered and discarded. On the other hand, a poorly implemented VM programme (for instance, the VM team not analysing the required functions properly) could make the study nothing more than a cost reduction process. Since the required functions may not be satisfied, a poor design may result.
2	Although the functional approach in the VM methodology provides a fresh look at the design problem, it is sometimes difficult to provide a holistic view of a complex artifact such as a building in functional terms in one go.
3	The implementation of recommendations suggested by a VM team depends, to a large extent, on the co-operation of the original design team. To get the designers actively involved and co-operative is vital to the success of any VM study. It is also one of the most difficult aspects of VM studies.
4	In theory, a VM programme can be implemented at any stage of a construction project. In practice, however, later applications of VM could cause a huge amount of abortive design work and delay to project completion. The VM change proposals are therefore unlikely to be accepted by the designers and clients.

5.4   Critical Success Factors

There are a number of factors that will affect the fate of VM studies. Successful studies do not come naturally. Based on interviews with construction professionals, the critical success factors for their application in Hong Kong's construction industry are as follows:

MANAGEMENT

1. Project Selection

The project selection, which affects the effectiveness of VM, is one of the critical success factors of VM studies. As shown in Figure 5.1, the cost of VM is minimal compared to the cost of a construction project. Generally it can be expected that the cost of VM will be considerably less than 1 percent of the total construction project cost. Since it can be expected that a fairly well administered VM programme should yield cost reductions in the order of 5–10 percent of construction project costs, it can be seen that VM literally pays for itself (Norton and McElligott, 1995).

2. Timing of Study

Traditionally, the VM technique is used at the detailed design stage, when most of the essential design decisions have been made. Because of its successful performance when applied to building design, this technique was developed and recommended for use throughout the design of a building project. In theory, VM can be applied at any stage of the building design process but many researchers and authors suggest that VM is usually applied to a project at the embryonic stage (i.e. conceptual design and sketch design stages) because the cost of making changes would increase over time during a project's life (Norton and McElligott, 1995; Kelly *et al.*, 2004)

Figure 5.2 show that the earlier in the design process that the study is undertaken, the higher the cost reduction potential and the lower the abortive design costs and costs of implementation. It can therefore be concluded that the maximum cost reduction potential occurs earlier in the briefing and design processes and therefore to delay VM application until the construction phase would severely limit its potential impact. However, clients usually initiate a VM study only when it becomes apparent that the cost of a project outweighs its worth and the design is over 70 percent complete. As a result, the value of VM is significantly reduced causing much disruption and abortive design costs.

3. Selection of Study Team Members

The selection of members for a VM study team will depend on the subject of study, with the number of participants typically varying from 4 to 20, and in some cases more than 20. It is necessary that the key individuals and participants are familiar with the process, are committed to its implementation, and know what to expect from the study. The team may comprise stakeholders, users, suppliers or multi-

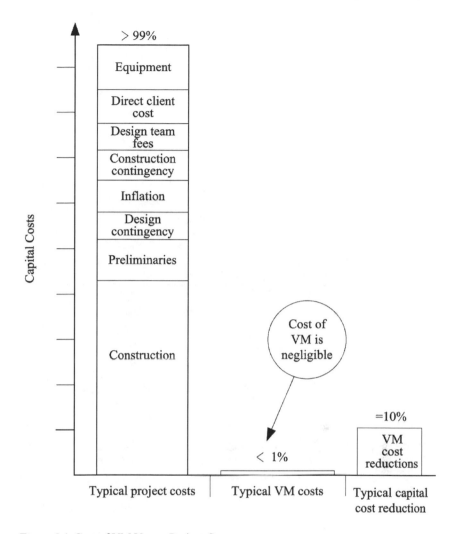

Figure 5.1 Cost of VM Versus Project Costs

Adapted from Norton and McElligott, 1995.

disciplinary specialists, with some being involved full-time and others possibly called in at an appropriate time during the study. Moreover, the composition of the team is also important as this may directly affect the study output. Hence, it is essential to select the most experienced available members in order to enhance the credibility of the recommendations produced (AS/NZS 4183:1994).

4. Support from Senior Management

Genuine support from top levels of management is important as this ensures timely resource approval, staff availability, and committed participation.

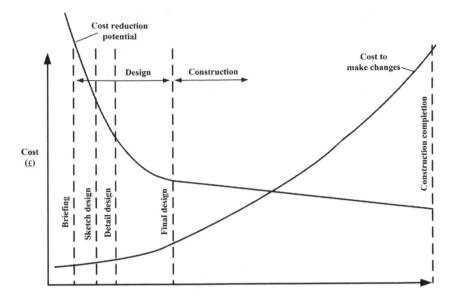

Figure 5.2 Cost Reduction Potential Versus Cost to Implement Changes
Adapted from Norton and McElligott, 1995.

WORKSHOP MANAGEMENT AND FACILITATION

1. Facilitation Skills

The facilitator plays a key role and is a significant factor in the degree of success. The facilitator must have methodical training in both VM methodology and team facilitation. The facilitator should be independent of the study team with whom the study is being conducted and should be able to promote a constructive team spirit, encourage the cooperation of all participants, and engender and maintain enthusiasm, all without unduly influencing decisions or dominating activities.

In order to perform this pivotal role, the facilitator must have a thorough knowledge of VM processes, procedures and techniques and should demonstrate competence in a wide range of skills that, according to AS/NZS 4183:1994, include:

- Group/team management skills
- Communication skills
- Analytical skills
- Interpretation skills
- Questioning skills
- Lateral thinking skills

2. Positive Human Dynamics

VM encourages positive behaviour, with all participants becoming involved in the full scope of the study and not limited to concern for their own sectional interests. The following should be encouraged:

- Positive and pro-active teamwork
- Recognition of individual contributions and team results
- Challenge of the status quo to bring about beneficial change
- Communication through shared understanding and group decision support seeking consensus
- Ownership of proposals by those responsible for implementing them

3. Consideration of the Organisation Environment

- Taking pre-existing external conditions into account, where organization managers may have little or no influence.
- Taking account of the influence that managers may have on external or internal factors that may impact the study.
- External or internal conditions influence the limits of the potential outcomes and should be quantified. These conditions can be either opportunities or constraints for the project.

> **Do you think that the factors described above are equally important to the success of VM applications? If not, which factor is the most important and what is your reason?**

5.5 Problems

Encountered Difficulties in VM Studies

The encountered difficulties in VM studies have been summarised into three major areas, which include (i) lack of information, (ii) lack of participation and interaction, and (iii) difficulty in conducting evaluation and analysis. These are presented in Table 5.2 followed by a detailed discussion.

1. Lack of Information

Research findings show that lack of information is ranked as the most frequently encountered difficulty in VM studies. This is in line with the work conducted by Park (1993), who suggested that insufficient information is a key problem in VM studies. Some reasons are suggested below.

Research findings show that lack of information is ranked as the most frequently encountered difficulty in VM studies. This is in line with the work conducted by Park (1993), who suggested that insufficient information is a key problem in VM studies. Some reasons are suggested below.

Table 5.2 Summary of Difficulties Encountered in VM Studies

Problem	Reason	Impact
1. Lack of information	Project information is poorly coordinated in the pre-study phase	Increases the 'uncertainty' in evaluation
	The difficulty of retrieving project information in VM workshops	
2. Lack of participation and interaction	Shy of public speaking Domination Poor team spirit Conformance pressure	Members' contributions of are reduced
3. Difficulty in conducting evaluation and analysis	Insufficient time to complete analysis Insufficient information to support analysis	Members are unable to respond to 'what if' type questions

Project Information is Poorly Coordinated in the Pre-study Phase

Respondents to a survey suggested that all project information should be gathered to produce a completed data file in the pre-study phase, but this process is not well coordinated in the construction industry. In Hong Kong's construction industry, the duration of a VM study has been shortened from 8–24 hours to 4–16 hours. Some argue that it is costly to arrange for 20–30 stakeholders to work together in meetings. Many clients are trying to shorten the duration of VM studies and thus, a one-day (8 hour) workshop seems to be the most popular in the construction industry. Also, as facilitators have difficulty in arranging pre-meeting workshops to coordinate information exchange and gathering, project information is reported to be poorly distributed and circulated in VM studies. In a worst-case scenario, participants only receive the project information paper and the set of documents describing the study, one day before the workshop. In addition, stakeholders from various departments or organisations invited to join the study teams must acquire project information prior to the studies. Consequently, they do not have sufficient time to study the paper and prepare the relevant materials, and hence, function analysis cannot be started quickly at the beginning of workshops.

The Difficulty of Retrieving Project Information in VM Workshops

Respondents also reported that the direction of creative thinking is unpredictable, and it is therefore difficult for them to ensure that all relevant project information is ready for workshop. Moreover, a conference room is a semi-closed environment and the physical boundary may prohibit them from retrieving any new information during workshops. As a result, respondents lack project information in VM studies.

They often make assumptions and put 'unanswerable questions' into an action plan. This will delay subsequent tasks and increase 'uncertainty' in the evaluation phase.

2. Lack of Participation and Interaction

The VM process is centred upon a participatory workshop involving a multi-disciplinary, representative group of people working together to seek the best value solution for a particular situation. Thus, the contributions and involvement of stakeholders are important to the success of VM studies (AS/ NZS 4183: 1994 and Reichling, 1995). However, survey respondents ranked lack of participation and interaction as one of the most frequently encountered difficulties and suggested that this problem is very common in VM studies. Some reasons for this are given below.

Shy of Public Speaking

Due to their personality, some members are shy of public speaking and are therefore reluctant to speak out in VM workshops.

Domination

The duration of the VM workshop is very short, so any domination of the discussion would result in an uneven chance for each member to participate. Respondents pointed out that the problem of domination is commonly found in VM studies because of the conflicting objectives or interests of the participants. A few active members may tend to dominate the discussion and prevent others from participating in the process. This may result in lower participation and interaction among members in VM studies.

Poor Team Spirit

The VM team is a temporary formal group. Members come from different disciplines and organisations and therefore need extra time to develop trust and good relationships so as to effectively integrate as a VM team. However, the short duration of VM studies imposes difficulty in team building, and some of them may therefore lack a sense of belonging and not fully contribute to the studies. As a result, 'free-riding' may occur in VM workshops.

Conformance Pressure

Conformance pressure from senior members of a VM study team may prohibit the interaction and participation of junior members. The members of the VM team come from different hierarchical levels, including senior executives, middle managers, and workers. Hence, there is a possibility that senior members may, intentionally or unintentionally, exert pressure on junior members For example, junior members may be afraid to criticise bad ideas from senior members because of the traditional culture and social status. As a result, some junior members become inactive and remain silent in VM workshops. To overcome this problem, facilitators apply various techniques to promote active participation. For example, they may make use of role-playing to motivate junior workshop members. However, the factors described above are believed to inhibit workshop members' participation and interaction to a large extent in VM studies.

3. Difficulty in Conducting Evaluation and Analysis

The difficulty in conducting data analysis is also ranked as one of the most frequently encountered difficulties in VM studies. The reasons are listed below.

Insufficient Time to Complete Analysis

Survey respondents suggested that the ideas produced in the creative phase require extensive consultation and in-depth investigation to analyse their feasibility and potential benefits, but that they have insufficient time to complete the analysis in VM workshops. Norton and McElligott (1995) suggested that the analysis activities, such as backup calculation and cost analysis, are time-consuming and might occupy half of the time of a VM workshop. In addition, the problem of time constraints is believed to be more serious in Hong Kong than in America and Australia. Shen (1997) suggested that the duration of VM workshops is very short and some sessions are used to educate participants who are unfamiliar with VM processes. As a result, it is difficult to complete all the necessary analysis within VM workshops in Hong Kong's construction industry.

Insufficient Information to Support Analysis

Apart from the problem of insufficient time, respondents also suggested that they had insufficient information to support analysis in VM studies, as discussed in previous section. Members cannot therefore conduct the evaluation and analysis processes efficiently in VM workshops, and sometimes they are unable to respond to the 'what if' type of question very quickly. This may delay the progress of the evaluation and development phases in VM studies.

> **Have you come across any difficulties similar to those above, in your meetings? If so, what are the reason(s) for those difficulties?**

5.6 References

Jones, J.C. (1970). *Design Methods: seeds of human futures*. London: John Wiley & Sons Ltd.

Kelly, J., Male, S., and Graham, D. (2004). *Value management of construction projects*. Oxford: Blackwell Science Ltd.

Norton, B.R. and McElligott, W.C. (1995). *Value Management in Construction: A Practical Guide*. Basingstoke: Macmillan Press.

Park, P.E. (1993). Creativity and Value Engineering Teams, *Proceedings of the 28th SAVE International Annual Conference*. Fort Lauderdale, FL: SAVE International.

Reichling, A.L. (1995). What Makes a Successful Team? *Proceedings of the 35th SAVE International Annual Conference*, 99-102.

Shen Q.P. (1997). A critical review of VM applications in the construction industry in Hong Kong, *Proceeding of the 30th Annual Conference of the Society of Japanese Value Engineering*, 221-226.

Standards Australia (1994). *Australian/New Zealand standard: value management* (AS/NZS 4183: 1994). Homebush: Standards Australia.

Zimmerman L.W. and Hart G.D. (1982). *Value engineering: practical approach for owners, designers, and contractors*. New York, NY: Van Nostrand Reinhold Company.

6 VM Applications

*Geoffrey Q.P. Shen, Ann T.W. Yu,
and Jacky K.H. Chung*

6.1 Introduction

This chapter introduces the development of VM in different countries, particularly Hong Kong and China. It illustrates current practices and discusses the obstacles to the application of VM in the construction industry.

6.2 Learning Objectives

Upon completion of this chapter, you should be able to:

1. Demonstrate an awareness of the development and application of VM in Hong Kong and China
2. Explain the obstacles of VM application and its future prospects in the construction industry in Hong Kong and China
3. Discuss the development of VM in European countries and understand the differences in VM practice between Hong Kong, Australia/UK, and USA

6.3 VM in Hong Kong

The potential benefits of VM and its application have increasingly been recognised in Hong Kong, and there have been an increasing number of VM studies in various sectors, especially in the construction industry. Over the past 20 years, Hong Kong has also provided a platform for different practices to demonstrate their merits and to compete with each other for this attractive market in the construction industry. In this chapter, development and application of VM in Hong Kong is described and the findings of an industry-wide survey on the applications of VM in Hong Kong's construction industry are presented. The key obstacles of VM application in the construction industry are reviewed and a number of high priority actions are proposed to meet the opportunities and challenges in the application of VM in Hong Kong.

VM Development in Hong Kong

> **Are you aware of the historical development and the latest developments of VM applications in Hong Kong's construction industry?**

Since VM was first introduced in Hong Kong in 1988 there has been an increasing awareness of its merits and tremendous potential in value enhancement and cost savings. In the two decades after its introduction, VM applications greatly increased as can be seen from the following examples.

Table 6.1 VM Studies in Hong Kong's Construction Industry

Project Name	Year
Value Engineering Training Workshop	1988
United Christian Hospital	1988
Hong Kong Cable Television Network	1989
South China Morning Post Building	1993
North District Hospital Project	1994
Various PAA Projects	1994
KCRC Development Project	1994
Lei Yue Mun Housing Project	1994
Stanley Prison	1994
Haven of Hope Hospital	1995
KCRC Western Rail Project	1996
Various projects in ArchSD	–
Various projects in Works Bureau	–

Many VM studies for construction projects have been initiated by the Architectural Services Department (ArchSD), one of Asia's largest multi-disciplinary professional offices in the public sector, which has played a leading role in promoting and using VM in Hong Kong. Table 6.2 shows some of the studies. The total value of these projects is around HK$14,150 million. The objective of the ArchSD is to introduce VM as a standard methodology in the initiation and implementation of all major projects (Wilson, 1996).

In May 1996, less than one year after its incorporation, the Hong Kong Institute of Value Management (HKIVM) successfully organised a two-day international conference, Value Management in the Pacific Rim, which attracted more than 100 speakers and attendants from the USA, UK, Australia, New Zealand, Canada, Brazil and Hong Kong. The Secretary for Works and the Chairman and Chief Executive of the Kowloon Canton Railway Corporation also addressed this conference.

Table 6.2 VM Studies Initiated by the ArchSD

No	Project Name	Value
1	Mongkok Stadium	250M
2	School Improvement Project	5000M
3	Shengshui Slaughter House	1208M
4	Ecological Park	300M
5	ICAC Headquarters	870M
6	Government Dockyard	150M
7	CAS & FSD Kowloon Resources Training Centre	256M
8	Wetland Park	430M
9	Swimming Pool Complex in Area 1 (Sun Wai Court), Tuen Mun	429M
10	Cremators at Wo Hop Shek Crematorium	335M
11	Joint-user Complex & Wholescale Fish Market in Area 44, Tuen Mun	238M
12	Customs Headquarters Building	1073M
13	Lo Wu Correctional Institution	1376M
14	Swimming Pool Complex in Area 2, Tung Chung	410M
15	Prince of Wales Hospital – Extension Block	1822M

Between 1996 and 2014, eleven international VM conferences have been successfully organised by HKIVM. It has also enjoyed a steady growth in membership and a wide membership base that consists not only of VM facilitators, but also academics and managers from client organisations.

Over the last twenty years, there have been numerous activities in Hong Kong aimed at promoting VM awareness and application. They include awareness seminars and learning workshops conducted at various professional institutions such as the Hong Kong Institute of Architects (HKIA), the Hong Kong Institute of Engineers (HKIE), the Hong Kong Institute of Surveyors (HKIS), Hong Kong Institute of Construction Managers, and the Chartered Institute of Building (CIOB) Hong Kong, as well as at some large commercial organisations such as the Mass Transit Railway Corporation (MTRC).

On the education front, since 1995, the Department of Building and Real Estate (BRE) of The Hong Kong Polytechnic University has offered 'Value Management in Construction and Property' as one of its core modules for its MSc award. The module consists of 24 contact hours including 8-hour participation in hands-on workshops. A similar subject 'Cost and Value Management' has also been offered to undergraduate students. This module consists of 42 contact hours including 3-hour participation in hands-on workshops.

A number of research projects have also been set up in order to further develop VM theories and improve their application. Some VM-related research projects undertaken at The Hong Kong Polytechnic University are listed in Table 6.3.

Table 6.3 Government Funding for VM Research

Project Title	Funded by	HK$
Successful value management applications in China's market economy: development of a framework for implementing the best practice	RGC Competitive Earmarked Research Grant	$380,000
Functional representation of client requirements for building projects: development of a briefing guide and a framework for knowledge-based implementation	RGC's FBN and funded by PolyU	$240,000
Development of web-based diagram-oriented group decision support system for project scope management	UGC Block Grant/ PolyU	$450,400
Benchmarking the best value management practice in China's construction industry	UGC Block Grant/ PolyU	$452,710
Bench-marking and development of a process model for applying VM in construction projects	CERG's FBN, funded by PolyU	$240,000
Applications of value management in the construction industry	RGC's direct allocation	$180,000
A best practice framework for systematic identification and precise representation of client's requirements in the briefing process	RGC CERG	$433,400
Effect of using group decision support systems on the processes and outcomes of value management studies	RGC CERG	$420,400
Measuring the processes and outcomes of value management studies in construction	RGC CERG	$353,537
A computer-aided toolkit for using the functional performance specification in the briefing process of construction	RGC CERG	$380,000
Managing multiple stakeholders in the briefing process of large construction projects	RGC CERG	$648,000
The effect of using group decision support systems on virtual value management workshops for major projects	RGC GRF	$644,700

VM APPLICATIONS IN HONG KONG'S CONSTRUCTION INDUSTRY

Before the Asian Financial Crisis in 1997

While the above section demonstrates the rapid development of VM in Hong Kong, especially in the public sector, the overall picture of current VM applications in the private sector of the construction industry is not very encouraging. There were not enough applications in the industry and this can be seen from the findings of a research project undertaken in 1996.

The main objective of the project was to investigate VM awareness and applications in Hong Kong's construction industry. A questionnaire survey was conducted in April 1996, with a total of 796 questionnaires sent to senior personnel of selected contractors, consulting firms and property developers in the private sector of the industry. As 75 valid responses were received a return rate of 9.4 percent was achieved. The respondents consisted of 42 contractors, 7 developers, 19 consultant firms, and 7 unidentified respondents.

According to the survey results, awareness of VM in the private sector of the construction industry in Hong Kong was low. Around 40 percent of the respondents had not heard of any of the key terms such as value management, value engineering, value analysis, value control, and function analysis. Among those who had heard of these terms, less than 12 percent had a high level of knowledge and understanding of VM and its related terms, others had limited knowledge.

Within the questionnaire, a number of questions were included to test whether the respondents had a proper understanding of VM and to reveal their perceptions about VM. Statements such as 'VM equals cost reduction', 'VM is the same as cost planning', and 'VM is a value enhancing tool rather than a cost cutting method' were used. Although the majority passed these tests, around 35 percent of respondents did not respond correctly. This reflects possible misconceptions about VM.

The survey also revealed that respondents' experience of VM applications in Hong Kong was relatively low. Among the 75 respondents, only 15 (20 percent) said they had participated in VM studies. As shown in Table 6.4, there are many reasons for not using VM, but the most frequently cited was the lack of knowledge to implement it.

The low level of applications was probably associated with the low level of awareness of VM among senior management in client organisations too. The survey showed that only around 15 percent of contractors and consultancy firms had been asked by their clients to conduct VM studies. This is understandable since it is very unlikely for clients who have little or no knowledge of VM to ask their designers and contractors to conduct VM studies for their projects.

Findings from the above research project indicate that although the awareness was rather low and the experience rather limited, the majority of respondents were very interested in the topic and were eager to learn more about it. Furthermore, it is reasonable to speculate that many of the firms that failed to respond to the survey had no knowledge of VM and may have been afraid of being identified as vulnerable or old-fashioned.

After the Asian Financial Crisis in 1997

The economic downturn in Asian countries had a significant negative impact on all business sectors. The construction and real estate industry, previously one of the powerhouses of the economy in Hong Kong, was no exception. Starting from late 1997, the industry has been experiencing a long and painful adjustment period, with property prices coming down by as much as 60 percent in certain developments. Many existing building projects were postponed and few new projects were started.

Table 6.4 Reasons for Not Using VM at Work

Reasons	Percentage
Lack of knowledge to implement this new approach at work	25.4
No confidence to introduce VM to clients	25.3
Lack of time to implement this new approach at work	21.4
The client and/or other team members are reluctant to change	17.3
Traditional cost saving methods are more adequate or better	9.3

As a result, fewer VM studies were undertaken for the construction and property industry. However, by drawing upon experiences in VM studies in Hong Kong and by referring to experiences from overseas, it can be concluded that the VM workshop is a useful tool during an economic downturn. This section gives a detailed analysis of the implication of the financial crisis on VM applications, and how VM services should be improved at such an opportune time.

Previous empirical studies on applications of VM in Hong Kong revealed its slow development after the Asian Financial Crisis in 1997 (e.g. Grosvenor, 1993; Shen, 1997a, 1997b; Fong *et al.*, 1998). Although VM was first introduced to Hong Kong's construction and property industry in 1988, its application was widely regarded as being in infancy. There were two fundamental reasons for this slow progress.

- Firstly, land in Hong Kong is extremely scarce and expensive: land price and its associated costs constitute around 70 percent of the total cost of a residential development. There was little incentive for developers to apply value management on the design and construction process of a development, which accounts for only 1/3 of the total development cost.
- Secondly, because of the strong economy and the booming property market, many property developers can enjoy very good profit margins in their residential development projects. Such good profit margins satisfied their business goals, and there was no incentive for them to apply value management in their project.

The economic situation in Hong Kong changed significantly due to the Asian financial turmoil. The reasons described above, which include high land costs and high profit margins, are less compelling in the light of the economic changes. What impact did the economic crisis have on the application of value management in the property and construction industry? The following section attempts to analyse the changes and explore their possible impact on applications of value management.

Before the financial crisis, large developers enjoyed very good profit margins; small and medium developers also made reasonably good profits in their piecemeal developments. However, this situation was changing rapidly due to the impact of the economic recession, and the changing environment in which this sector was operating, e.g., the housing policies of the Hong Kong Government and its determination to maintain sustainable and healthy development in the housing sector.

The rivalry among property developers in Hong Kong was becoming increasingly intense due to the sluggish economic growth and reduced demand in the residential sector of the industry. There was fierce competition among developers. In addition, the private housing sector faced a potentially increasing threat from its growing substitutes, as more and more good quality public housing in desirable locations were made available to the general public for purchasing and renting. These changes had serious implications for VM applications in the real estate and construction industry.

Because of the significant drop in housing prices, the cost of land constituted a smaller portion in the total development cost than before. As shown in Figure 6.1, we assume that in the initial Scenario 1 given below, the proportion of costs for building and construction (B&C) and land is 1:2. If the average housing price drops by 33 percent, the cost for B&C is likely to remain unchanged and the new ratio between the costs for B&C and land will be 1:1 as in Scenario 2.

In several land sales offered by the Government after the Asian financial crisis, sale prices were halved in comparison with previous sales. The increasing proportion of construction costs provided more incentive for developers to consider the use of VM for their projects.

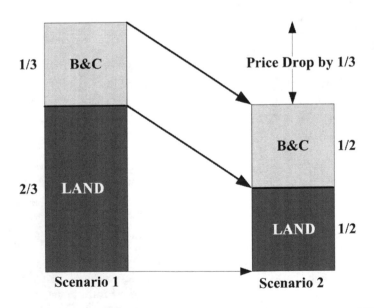

Figure 6.1 Changes in Cost Proportions

The housing sector was affected by high interest rates and financial crises in neighbouring countries. Because of the declining economy, rising unemployment, high local interest rates, and a sluggish retail sales market, there was intense competition among developers, and the profitability levels of the housing sector was further reduced. Property developers in Hong Kong were not able to enjoy the high profit margins as before the financial crisis. They therefore exercised tight control over costs, empowering themselves with useful tools such as VM, Building Information Modelling (BIM) and information technology to enhance the value of their projects and compete in volume.

A VM study was conducted for a public housing project in Tuen Mun after the scheme design was completed. The purpose of the study was to:

- Review the project definition in the light of the rapidly changing environment
- Ensure that the project was defined properly
- Investigate alternative solutions to achieve the functions required by the client.

Whereas the changing environment might bring opportunities for more VM applications, it might also bring a threat to the development of VM. As many business entities in Hong Kong were in a cost-cutting mode in order to become more competitive in the marketplace, they did not feel it worthwhile to spend money on VM studies.

Current situation

The increase in materials and labour costs in recent years has created a very good opportunity for VM's further development in Hong Kong. This opportune time is further reinforced by the following events:

- First, the Development Bureau Technical Circular (Works) No. 35/2002 demonstrates that the Government is determined to enhance value for money by applying VM in all major public projects over HK$200 million. This leading position assumed by the Government undoubtedly has a positive impact on VM in Hong Kong.
- Second, the certification procedures adopted by the HKIVM to recognise Value Management Facilitators (VMF) also help to ensure the quality of VM studies, which provides additional confidence to client organisations that are planning to use VM in their projects.
- Third, the high demand in housing supply and the substantial increase in construction cost help to promote the concepts of value for money and the value methodology. During this time, people start to think of ways to enhance value and to save costs.

VM studies can be used in many ways in the real estate sector to improve business both in the short and long term. These studies can be strategic in nature by looking at possible future portfolios or the way a company will operate in the years to come, or they can be more tactical by focusing on near term individual development.

In its application to individual developments, VM studies can be usefully applied at many key stages within the development process. A series of workshops can be planned, each focusing on the decision-making process at the corresponding stage of the development process. Different studies can also focus on different aspects of the development.

What are the strategic issues that can be addressed through a VM Workshop? A VM study can be used to develop a consensus on the planning changes needed to meet community expectations in respect of the size and quality of accommodation. The study may also be extended to include:

- Experts in home ownership
- Likely trends in public and private rental costs
- Possible trends in earning and purchasing capacity following a downturn in the market
- Recovery as well as the impact of the physical changes to Hong Kong and the impact on end-users

VM studies must dare to look into the future and to consider different scenarios and their impact on their business. A VM study provides an ideal forum for developing the strategic direction of an organisation. Individual property developers will have their own unique focus and need to consider where the market is heading and what strategic business plans they need to plan and implement.

The information phase of the VM study should include identification of market segments, threats and opportunities, and corporate objectives, all of which should be developed and discussed. This would continue in the analysis phase in order to evaluate and prioritise objectives, risks and concerns. Finally, the creative, judgement and development phase would then synthesise options and establish an agreed, clear strategic direction for the developer.

The increasing significance of construction cost in the total cost of development provides developers with the incentive to seek out a wider range of options to meet the newly defined market. Where possible they may redesign their buildings so that the units respond to this new market, which seeks to provide housing at different and changing levels of affordability. Past trends in end-user expectations should be reassessed. This may mean not only changes in quality but also in the size of units – smaller for single families or larger for extended families. Is accommodation for live-in help still required?

In addition, VM studies can be applied to improve procurement strategies, project management methods and construction methods (such as more extensive use of precast concrete) in construction. At project level, a study provides the opportunity to reconsider project objectives and assumptions that underpin both the real estate and construction sectors, and to search for new solutions.

The Government, as Hong Kong's largest landlord, may also usefully reflect on its approach to housing, both for the short and long term. The Government spends heavily on subsidised public housing. They must continue to review how it can continue to provide its services more efficiently and effectively. Numerous initiatives are underway in this area. A VM study provides a means of harnessing the creativity and support of a wide group of relevant stakeholders in addressing this major programme.

The VM workshop is just one tool in the real estate managers' toolbox. It is a very powerful yet underused tool, which with greater awareness and understanding may be adopted by both the private and public sectors at this critical point in Hong Kong.

The recent and high demand in the residential property market should lead to wider application of VM workshops in the industry. It is hoped that proponents of VM can create a wider awareness of the benefits of effective VM studies for client organisations, and that there will be a surge in the number of VM studies in the real estate industry. As competition within the industry becomes increasingly intense, more and more developers are searching for useful tools such as VM to help them remain competitive in the market place.

There has been a surge of interest, especially in the construction industry, over the past few years. Looking at the road ahead, there are plenty of opportunities as well as challenges. Experiences in many other countries show that the prosperity and development of VM applications rely heavily on Government support. Hong Kong is in a fortunate position, as several Government departments have committed themselves to increased application of the methodology to the initiation and implementation of projects in the public works program. Opportunities must be seized and proper actions taken if VM is to be applied widely and successfully in Hong Kong's construction industry.

Obstacles of VM Application in the Construction Industry

> **What do you think are the key obstacles of VM application in the construction industry?**

High Land Cost

> Land for buildings in Hong Kong is very scarce and Government land is usually allocated to the highest bidder. Due to the rise in demand for property, the price of land is driven to an extremely high level and, as a result, construction costs are just a minor part of the total development costs
>
> (Fong and Shen, 2000)

Many would ask the question why should we bother to apply value management to save just a small percentage of that minor part? (Fong *et al.*, 1998).

Thus, even though there are significant benefits in applying VM, it is not surprising that most clients in Hong Kong show no enthusiasm for adopting it.

> The most important objective of most clients is releasing the property to the market as soon as possible, such that the investment can be repaid and generate profit, and they fear that VM studies would prolong the design period and adversely affect their investment plans
>
> (Fong and Shen, 2000)

In the current climate, it is hard for consultants and construction professionals to promote VM and explain the process of VM application in projects to their clients. The survey shows that most professionals seldom have the chance to practice VM, and because of the lack of demand from the clients for applying VM, most consultants are not allocating resources to VM services (Fong and Shen, 2000).

Procurement System

> In Hong Kong, a traditional system has been adopted for many years in procuring various types of building projects, and it is more popular than other options such as design and build (D&B), management contracting (MC) or build, operate and transfer (BOT)
>
> (Fong and Shen, 2000)

Even though clients and professionals often apply the system to construction projects, it does not mean the system is trouble-free. The main complaints are on the price uncertainty, long completion times, and lack of buildability. Despite this, the traditional system is still the preferred procurement method in Hong Kong. Ho (1995) argued 'most local firms are already familiar with and accustomed to the traditional system, and there is resistance to change', which is the reason for the preference.

Because of the clear separation of the design and construction phases in the traditional system, it is hard for contractors to contribute their expertise and technical knowledge in the design process. This is disadvantageous to a VM study where the presence of the contractors is a vital ingredient to improving the design and ensuring a high degree of buildability.

Professionalism

When conducting a VM workshop, the original design team, an external team, or a hybrid team takes the lead as the VM team. When an external team of professionals are involved in design review exercises, it can be expected that the original designers will be defensive of their own design, especially if the objectives of the study are not yet clarified and there is no appropriate briefing. Designers are often critical of VM studies: some oppose the implementation of a VM study because it is costly and time-consuming, and some question the qualifications of the team members (Fong and Shen, 2000).

Further arguments, such as the responsibility for design liability and who should bear the cost of re-design, further aggravate the relationship between the designer and the VM team members. Sometimes, the discussion is too emotional to be objective, resulting in a non-productive debate rather than a constructive study

(Fong and Shen, 2000)

As stated by Kelly and Male (1988):

The very nature of the VM process, dynamism, a change orientation and being perceived as potentially conflict laden, means that it could be seen as undermining traditional professional practices and procedures rather than being complimentary.

Kelly and Male (1988)

Problems Implementing the SAVE 40-hour Job Plan

The SAVE 40-hour VM Job Plan is widely-used in many VM studies world-wide and is regarded as a milestone by SAVE International and many VM organisations in other countries. This job plan has proved to be successful over the past four decades by many practitioners. If well organised, it can produce excellent results for minimum effort. However, it is very unlikely to be accepted and applied in the construction industry in Hong Kong.

The main problem with implementing the SAVE 40-hour Job Plan, is time. It is normally difficult to assemble key project participants for such a concentrated period of five days and retain their undivided attention throughout this period (Kelly and Male, 1991). This is especially true in Hong Kong, where because of the sky-high land prices, many clients have to pay a huge amount of interest for the money borrowed. They therefore normally give the designers and other consulting firms a very limited period of time to complete the design and other related works. Another reason is cost: because of the clients' limited experience, they are unlikely to endorse a study that will cost a relatively large amount of money, due to Hong Kong's high consultancy fees.

A two-day workshop seems to be too short to complete the necessary analyses, evaluation and development of alternative solutions however. Since many sessions must be devoted to educating participants who are rarely familiar with VM processes and principles, it is rather difficult for the VM team to work efficiently on the problem in hand. In addition, the evaluation and development of alternative solutions are particularly difficult to complete effectively in such a short time, because many ideas proposed in the creativity phase require intensive design and engineering analysis, particularly when these involve long-term life cycle cost trade-offs (Kelly and Male, 1991). Occasionally, the phases of information assimilation, development and presentation take too high a proportion of the duration of the workshop, and the time allocated to functional analysis and creativity is too short.

These drawbacks mainly relate to the time required for VM studies and the time allocation at workshops. The best way to address these problems seems to be an improvement in the efficiency of VM studies. With the assistance of modern information technology such as knowledge-based systems and BIM, the amount of time allocated for tasks such as the retrieval of information, the generation and evaluation of alternative solutions, and the presentation of study proposals can be reduced considerably. More time can therefore be assigned to more important tasks such as function analysis and the development of alternative solutions.

Future Prospects of VM in the Construction Industry

Since the birth of value analysis (VA) in the early part of the twentieth century, all VM tools aim at increasing the value of a VA subject, which is defined as the relationship between the satisfaction of needs and the use of resources in achieving this satisfaction. A mistake often made in the past, was seeing VA and subsequently VM as a cost reduction tool. Today, that is no longer possible because any VM study must take different stakeholders' expectations into account. The technical and economic aspects, together with the environmental and social aspects, form a complete package in which sustainability must also be considered.

The sustainability agenda in the construction industry has been growing steadily for many years. To ensure the most significant impact, sustainability must be included as early as possible in the briefing and design stages of a project, so that influence over decisions can be exerted. Sustainability must be a part of all activities within the project development process, and there must be close inter-action between clients and stakeholders. The progress towards more sustainable buildings is hindered by many barriers, such as the belief that internal trade-offs between the principles of sustainability must be made, the lack of guidance, a negative evaluation of green buildings in terms of additional cost, and a belief that sustainability is a separate problem.

VM can be used to promote sustainable design and development in the life cycle of a construction project. VM is a proactive, creative, problem-solving and problem-seeking method that seeks to maximise the functional value of a project by managing its development from concept to final use. Usage of VM could raise awareness of sustainability issues at the right time to eliminate unnecessary costs and integrate sustainability into the project, while staying within the budget. Using the systematic job plan, participants can be guided to consider sustainability issues throughout the life cycle of the building.

Only a few academic papers highlight the contribution of VM to the field of green buildings. Conferences by the Institute of Value Management Australia in 2002 and Hong Kong in 2008 presented several papers linking VM with sustain-ability, discussing the importance of sustainability as a tool to improve value (Barton 2002; Barton *et al.*, 2002; Schneider, 1999) and the potential of using VM to promote sustainability (Yeomans, 2002; Barton *et al.*, 1999; Philips, 1999). Some papers also call for a better integration of VM and sustainability so as to enhance the reputation of VM (Fong, 2003; Schneider, 1999). Research has shown that one of the barriers to implementation of green specifications in construction projects is 'the perceived additional cost' (Lam *et al.*, 2009), along with other factors such as the lack of guidelines (Lam *et al.*, 2010).

Despite this, using VM as a sustainability promoter in the construction industry has not been popular locally, nationally or internationally. This is unsettling as the lack of information in this area gives of an impression that the VM community and researchers are unaware of the importance of integrating VM and green building issues, and of the potential VM has in promoting sustainability. Integrating VM and sustainability would be a great way to address this problem, along with developing practical guidelines to be used during VM workshops to address green building issues.

There are examples of sustainability issues being incorporated in VM workshops; however, there are still gaps in the practice most likely caused by a lack of understanding and a lack of confidence in this practice (Abidin and Pasquire, 2007). The attention that sustainability issues are given is less than expected, and it varies between VM workshops depending on the importance that the client places upon it. Though many VM practitioners claim to have a strong knowledge of sustainability issues, and to be aware of its importance, the lack of absorption of this knowledge into the VM process shows that there is a need to use this knowledge and address green issues more effectively. By increasing awareness and understanding of the sustainability concept in the pre-workshop phase of a VM study, an improvement in sustainability can be achieved. Green building issues should not be treated as a separate agenda as they can form an integral part of the project functions and objectives, as such they should be dealt with during the VM process. To achieve this, the issues should be raised early in the process and be backed up by client commitment. Having some formal guidance available would be helpful to pinpoint where improvement initiatives can be focused. To help with this, research needs to be undertaken into strategies for a better integration of VM and green building design and construction. This will increase the likelihood of constructing high performance buildings with high value for money in the future.

6.4 VM in China

VM Development in China

In Mainland China VM is often referred to as Value Engineering (VE). In 1978, VE was introduced to China as one of the 18 modern management techniques as part of the country's economic reforms. While VE originated in the US, Japan was the major source of information for China.

One of the cities to introduce VE was Shanghai, where it came in the early 1980's and has been widely used since. In 1981, the first exhibition on VE results was organised by the Economic Commission of the Shanghai Municipal Government. In 1984, the same commission published its 'Working Plan for Expediting Value Engineering Applications', which required organisations across all industries to apply and use VE. Many institutions in Shanghai, among them the Shanghai Scientific Commission and the Shanghai Association for Science and Technology, strongly supported this.

Between 1978 and 1986, the Shanghai Value Engineering Society (SVES), in conjunction with other relevant institutions e.g. the Shanghai TV station, organised training programmes for more than 100,000 people (Zhu, 1988). Also during that time a sample survey of 580 VE studies conducted by SVES members, showed the total savings derived from VE studies were RMB 250 million. Among these studies, 79 applications were in new product development, 367 applications were in renovation of old products and 30 were used in software development. VE has also been successfully used in many other provinces and cities in China, such as Beijing, Tianjin, Guangdong, Shandong, Liaoning, Sichuan, Guangxi, Jiangsu, Zhejiang, Fujian, and Jilin.

Applications of VE can be seen in a variety of industries. These include machinery, textiles, agriculture, finance, coals and mines, defence, electronics, commerce, services, and many others. As shown in Table 4, a number of VE societies and associations have been established in various cities and industries. According to a report published in China Value Engineering (1990), the annual savings in China were around RMB 500million in the late 1980's.

The National Standard Bureau in China developed the first national standard on Value Engineering in 1987 (Value Engineering – General Terms and Work Program GB8223-87). The standard defined the terms 'Value Engineering', 'Value' and 'Function', and the purpose and characteristics of VE were introduced. It also featured the procedures that VE workshops should follow. The Standard was made effective from the 1st of July 1988.

In 1988 when the application of VE entered its 10th year in China, President Jiang Ze Min (who was Party Secretary in Shanghai at that time) wrote to the Shanghai Value Engineering Society. Praising the success of applying VE in Shanghai and the rest of China, he emphasised: 'the more we use value engineering, the more we can benefit from it.' The importance of VE application in China was also emphasised by the President of the China Association for Business Administration, who believed that VE could enhance the competitiveness of enterprises. In 1984 and 1989, he called for a full implementation of VE across China.

In 1992, the China Association of Business Administration, in collaboration with the China Association for Science and Technology, and the China Central Television Station, organised a 625-minute TV training programme on VE, which was the largest ever in the history of VE world-wide, attended by millions of people in China. A summary of the major historical events of VE in China is given in Table 6.5.

VE Research and Teaching in China

The VE Division of the China Association of Higher Education was established in August 1987. Starting in 1988, this institution has organised a biannual Value Engineering Symposium. The current members of the division include 78 tertiary institutions from all over China. Many national key universities such as Tsinghua University, People's University, Shanghai Jiaotong University, Tongji University, and Zhejiang University are members of the VE Branch.

There are over 50 VE books written in Chinese by professionals in academic institutions and various industries. There are also a number of textbooks translated

Table 6.5 Major Historical Events of VE in China

Month/Year	Major Historical Events
6/1978	Professor Shengbai Shen at the Shanghai Philosophy & Society Association first introduced VE with a presentation entitled 'Introduction of Value Engineering'.
12/1978	The first formal article introducing VE 'The applications of value analysis in Japan's automobile industry' written by Junbo Dai was published in the journal of Information of Foreign Manufacturing Industry.
8/1981	The Ministry of Manufacturing called for wider use of VE in the industry.
5/1984	The Achievements of VE Applications Exhibition was held in Shanghai and received high interest and recognition from those attending.
2/1985	The authorized formal journal 'Value Engineering' was first published.
9/1985	Yixin Shen, a member of the Chinese People's Consultative Committee (similar to Senate in the U.S), submitted a proposal suggesting that the Government promote the use of VE in China.
7/1987	The first national standard (GB8223–87) was issued by the National Standards Bureau.
3/1988	President Jiang Zeming wrote to the Shanghai VE Society 'The more we use VE, the more we can develop and benefit from it'.
8/1988	The VE Branch of the China Association for Tertiary Institutions was founded. It has since become an influential force promoting VE in academia.
11/1990	VE was included in the 'Eight Five' key training programmes by the China Science and Technology Association, the China Enterprise Management Association, and the China Entrepreneurs Association.
7/1992	A 2-month TV training program of VE jointly organised by the China Science and the Technology Association and the VE Branch of the China Enterprise Management Association was broadcast on CCTV.
12/1998	The First National VE Representatives Conference was held in Beijing.
12/1998	The Chinese Society of Value Engineering (preparation committee) was founded in Beijing.
5/1999	The Value Engineering & Technology Innovation International Conference was held in Hangzhou and was well received by local and overseas representatives.
11/2000	The VE committee of the China Manufacturing Enterprises association was formed.
11/2001	The China Manufacturing VE Society organized the Annual Conference in Beihai with the theme of ' Entry WTO, Use VE'.

Table 6.6 VE Related Institutions in China

Year	Name of Institution
05,1981	VE Branch of Management Modernisation Division, China Association for Textile Enterprises
12,1981	VE Branch, Hebei Institute for Economics and Management Modernisation
03,1982	VE Branch, Management Modernisation Institute, China Mechanical Engineering Association
07,1983	VE Branch, Ningxia Hui Autonomous Region's Association of Business Administration
10,1983	VE Branch, Sichun Association of Management in the Machinery Industry
06,1984	VE Institute for Agricultural Enterprises
10,1987	VE Branch, Guangdong Association for Business Administration in the Machinery Industry
12,1987	Shanghai Value Engineering Society
05,1988	Value Engineering Institute of the China Association for Business Administration
08,1988	Value Engineering Society for Tertiary Institutions

from either English or Japanese. As shown in Table 6.7, the number of VE related regular publications is high in comparison with other countries (although some of them may have already ceased publication). A large number of academics are undertaking research projects in the field of VE and it is taught in many universities in China.

Table 6.7 VE Related Publications in China

Title of Publication	Organisations Involved in the Publications
Value Engineering	VE Division of The China Association for Higher Education, Value Engineering Institute of The China Association for Business Administration
Value Engineering Express	The VE Institute of China Association for Business Administration
China Value Engineering	The VE Institute of China Association for Business Administration
VE Communications	The Guangdong VE Institute of Higher Education
Shanghai Value Engineering	The Shanghai Value Engineering Society
Value Engineering Brief	Division of Enterprises of Tianjin Economic Commission
Value Engineering Trends	Division of Jiangxi Association for Business Administration
VE Newsletter	VE Division of Shandong Association for Business Administration

VM Applications in the Industries

Through cooperation with Japanese manufacturers, VM was introduced to the Chinese manufacturing industry in 1978 (Han, 1998). Due to its success in the manufacturing industry, and advocacy by the Chinese Central Government, VM experienced significant growth in the following 15 years and was gradually adopted into other industries.

Due to China's transition from a planned economy to a market economy, the application of VM declined sharply in the mid-1990s. State-owned companies were the main users of VM in China, and with the radical transformation forced on them they were in a difficult situation. The established VM practices used under the planned economy were no longer viable, and no new strategy for using VM in the new economic environment had been established. It is still widely believed by local experts, that the new economic system will encourage and motivate a higher application of VM in the long term (Hu, 1999; Ma, 2000).

Shen and Lui (2004) carried out a survey to discover the state of VM practice in China, and to identify challenges to its further development. Two thousand questionnaires were distributed to chief engineers, one thousand based in manufacturing companies and one thousand based in construction companies (including design institutes and contractors). The findings are presented below.

Awareness of VM

According to the survey, 79% of the respondents from construction and 98% from manufacturing had heard of the term VM or other synonymous terms such as VE, value analysis and function analysis. The high awareness level in the manufacturing industry reflects that it is the most active field for VM usage in China, and VM knowledge has been widely disseminated in this sector in the past 20 years. However, it should be pointed out that actual awareness might be somewhat lower than these figures, since some respondents who had not heard of these terms might not have bothered to return their questionnaires

(Liu and Shen, 2005)

Areas of VM Application

The survey results show that only 24% of the respondents in construction had experience in applying VM, while the percentage was 51% in manufacturing. Among the respondents in construction, a small proportion indicated that they had never participated in a structured VM study, but that the philosophy and techniques of VM had been used to tackle problems encountered at work. VM activities in most manufacturing companies are based on the internal VM policy of the companies

(Liu and Shen, 2005)

A typical framework for VM implementation in manufacturing companies in China is illustrated in the following figure.

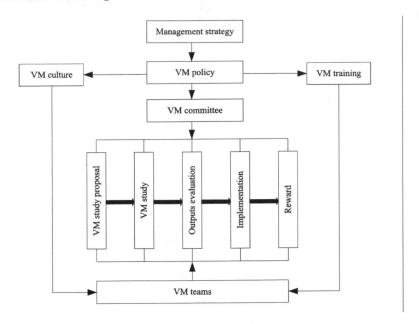

Figure 6.2 Typical Framework for VM Applications in China

Adapted from Liu and Shen, 2005.

Structure of VM Applications

A VM committee, often consisting of senior managers, engineers, financial staff and VM coordinators, is formed to manage, support and supervise VM activities

(Liu and Shen, 2005)

Normally, a VM study goes through the stages demonstrated in Table 6.8 below.

The survey results indicate that VM applications in China are confined to the manufacturing industry and rarely appear in other industries. Xiao (1998), found that over 80 percent of VM activities in China are implemented in the field of manufacturing. Because of this, most VM literature is related to the manufacturing industry, and local VM institutes in China are mostly represented by members from the manufacturing sector. There has been little effort to extend the use of VM to other industries. Where VM has been a large part of the construction industry for many years in Western countries, and continues to be one of the most active fields of VM application, the use of VM in China's construction industry is still in its infancy. To ensure the continued growth of VM in China, much more effort needs to be given to spreading the application of VM.

The design team or senior management of design institutes, rather than the clients themselves, initiate most VM applications in the construction industry. The tight budget that projects are normally under is an important reason driving designers to applying VM. Some contractors also create VM policies to encourage

Table 6.8 Three Stages of VM Application in China

Stage		Descriptions
1.	VM study proposal	Any department or temporary team within the company can submit a VM study proposal to the VM committee. In a proposal, the proposers explain the objectives, reasons, costs, schedule, estimated outputs, and the preparation for the proposed VM study. The VM committee will evaluate the proposal against a number of criteria. If it is approved, the proposed VM study will be conducted by a team of the proposers, under the management of the VM committee.
2.	VM study	The VM team carries out the VM study in accordance with the VM philosophy, job plan, and techniques. However, instead of a concentrated, continuous workshop such as a 40-hour workshop, the VM team members meet irregularly to explore, develop and test alternative solutions without suspending their normal work duties. It is therefore common for a VM study to last several months.
3.	Evaluation and implementation	The developed alternative solutions together with test reports will be submitted to the VM committee. Their effects on cost reduction and value enhancement will be thoroughly evaluated before they are fully implemented. If significant improvement is made from the VM study, the VM team will be rewarded according to the savings produced.

the use of VM in the construction stage, although this is mainly to optimize construction methods and create savings for the contractors, and thus has nothing to do with clients or savings during a project's lifecycle.

> Overseas experiences indicate that the mandatory VM requirements initiated by clients have made an important contribution to promoting VM applications, especially in the initial development stage of VM
>
> (Dell'Isola, 1982)

Educating clients seem to be a critical task for promoting VM applications in China's construction industry.

Differences in the Practice

From discussions with VM practitioners and experts in China, it can be concluded that the practice of VM in China is different from the overseas practice. The major differences are summarised in Table 6.9.

Table 6.9 Comparison of VM Practice Between China and Overseas

Items	Chinese practice	Overseas practice
Subjects of VM studies	Existing products/projects, often related to tactical problems.	Proposed and/or existing products/ projects, related to both tactical and strategic problems.
Facilitator of VM studies	In-house VM directors or engineers.	Internal or external VM experts.
Timing of VM studies	In design, production or construction stages.	From the concept to completion of a project/product.
VM team composition	In-house staff, several people familiar with the subject are involved.	Relevant stakeholders, often a large number of persons are involved.
VM workshop style	Informal workshop adopted.	Concentrated, continuous workshops preferred.
Function analysis	Using mathematical methods extensively to measure functions and to identify poor value.	The purpose here is to clarify clients' requirements and to understand their value system and identify poor value.
Duration	Depending on the subject under study, possibly several months	Normally lasting for only a few days.

Are there any similarities or differences between VM applications in Hong Kong's construction industry and in the construction industry in China?

Obstacles to VM Applications in China

Survey results suggest that VM applications in China are subject to difficulties, including lack of VM knowledge, lack of national VM standards, and lack of qualified VM facilitators. A summary of difficulties in applying VM in China is shown in Table 6.10.

Moreover, the survey also suggests that the most significant challenges in VM development in China are the limited scope of applications, the techniques used in the studies, and the measures adopted to promote applications. A brief discussion is given below.

The use of mathematical techniques at the function analysis stage has been advocated and emphasized by many influential Chinese VM authors. These techniques include value index, value graph, and pair-wise comparison (Li, 1998; Zou, 1998) and they often contribute to a significant portion of the Chinese VM books (Liu, 1998; Tang and Yang, 1996; Song, 1994). The most important objective of these techniques is to identify poor value areas in a product or project.

Talks with the VM practitioners indicate that it is the over-emphasis on the use of these techniques that has restricted the practical implementation of VM. This is

Table 6.10 Difficulties in Applying VM in China

Difficulties	Rank
Lack of national VM standards	1
Lack of VM knowledge	2
Lack of qualified VM facilitators	3
Insufficient time to carry out VM	4
VM prolongs product/project completion time	5
Too expensive to carry out VM	5
Defensive attitude of other professional teams	5
Interruption to normal work schedule	8

primarily because these techniques are developed by academics and are too complicated to be used in the industry, and secondly, the techniques are based on hard systems that focus on the components of a product or project. The underlying assumption in these techniques is that the value system already in place is correct for the corresponding product or project. This leads to the VM practitioners paying most of their attention to tactical issues rather than strategic issues. This is a problem because it keeps VM from being used to solve high-level problems in China.

> Government support made a great contribution to VM development in the late 1980s in China
>
> (Shen, 1997)

This promotion based on administrative measures also led to a misunderstanding of VM among some of the users. In the planned economy, Government administrative measures had a significant influence on the business of companies. As a result, to respond to Government calls for VM usage, some companies who had no training in VM, and had no experience using VM methodologies, simply labelled their cost-cutting exercise as VM to satisfy the Government. This practice has led to some companies wrongly believing that VM is just a new name for traditional cost control.

As has been the case in many countries, VM in China has not had a smooth progress. In the early 2000's VM applications declined measurably. Many of the local VM branches organized next to no activities. It is believed to be a result of the switch from a planned economy to a market economy. However, this is contrary to the theory that the new economic system should stimulate the use of VM in China.

As many companies are still coming to terms with the market economy, many of the state-owned enterprises are struggling to cope in the new business environment. Most people agree that in the long run, the development of VM in China and resulting applications will benefit from the new economic system.

However, the following difficulties are believed to be obstacles to the development and applications of VM in China.

- Lack of communication with the outside world
- Development of VM theories divorced from practice
- Non-existence of a national VM association
- Lack of certification procedures and professional recognition
- No research into the existing nation-wide application problems
- Poor knowledge transfer between sub-contractors

Future Prospects of VM Development in China

Looking towards the future, the prospects for VM in China are good. The change in economic system has brought with it many opportunities as well as challenges for state-owned and collectively-owned enterprises in China. One of the new driving forces in the economic environment is the entry of China into the World Trade Organization (WTO). To survive this new and globally competitive environment, Chinese enterprises have to improve and build on their competitive advantage. To do this, they are expected to turn to modern management techniques, including VM. The reorganisation of state-owned enterprises also benefits from the use of VM and other management techniques (Shen and Liu, 2004).

The year 1998 marked the 20th anniversary of VM applications in Mainland China, and as ever more economic reforms take place in the country, it is anticipated that state-owned enterprises will have fewer constraints in running their own business, which can lead to a higher degree of VM usage as companies look to improve their business. VM could prove a very useful tool for many of the company managers to learn and apply at work. There are strong reasons to believe that the next surge in VM applications in Mainland China is not far off. However, while some nearby countries have set up their own national value management organizations, China still lacks a unified and national organization that together with Government support can help promote VM throughout China.

Recommended Action

In order to revitalise VM activities, a number of actions must be taken. Referring to other countries' successful experiences with VM applications, the following issues should be addressed.

1. Development of theories must be linked with Practice

The last two decades has seen a great deal of research in this field, and many new VM theories have been put forward. They are valuable and significant contributions, as VM has not been used as much as other management techniques such as total quality management (TQM). However, most VM research has not been used in practice because VM researchers in academic institutions have little or no contact with those who are working with VM in the industry, which means that their findings are often too complicated or abstract to be of practical use in the industry. Another factor is the lack of promotion of new techniques in the industry, thus the research findings never gain popular usage. This shows that there is a very poor knowledge transfer between the academic development of VM and the industry application of VM. Academic researchers have a responsibility to ensure that their research findings are realistic and practical enough to be implemented and used in practice.

2. National certification procedures should be set up

In those countries where VM is widely recognised and used, there is a national VM institution whose main objectives are to promote VM, to maintain VM standards, and to enhance professionalism. A major tool adopted to achieve these objectives is Certification Procedures. In China, although there are some VM branches in a number of associations and institutions, there is no single organisation such as SAVE International, IVE in the UK, or the HKIVM in HK. Certification procedures hardly exist in any of those VM branches, and as a result VM lacks coherence between companies, cities and provinces. It is time to consider setting up a national VM institution that would take responsibility for establishing certification procedures to safeguard the interests of VM practitioners and users. A national VM organization would also be useful for handling the education and training of VM facilitators, and for keeping a database of approved and certified facilitators for companies to use.

3. Government support is vital to VM applications

As mentioned earlier, VM has benefited from initially obtaining Government support for national use. This support must be maintained if not enhanced. Members of the SAVE International are lobbying Congress in the USA, and IVE members are lobbying MPs in the UK. As the highest authority in China, the People's Congress should be persuaded to lend its support to the use of VM. This could also be an effective way to expand the use of VM in the construction industry, as the Government can require VM usage in all public construction works.

4. VM culture should be developed in business entities

During the introduction of VM in China, the main drivers were the local companies. In the late 1980's, these companies developed procedures for the full implementation of VM in their companies. This was partly due to the administrative measures dictated by the Central Government and local governments, and also as a direct result of the planned economy that was in place at that time.

With today's market economy, the relationship between local governments and the business entities in their jurisdiction has changed drastically. It is now up to the individual companies to decide what management techniques they use to enhance their competitiveness. The market economy promotes a conscious use of VM at all levels of business administration, which should eventually lead to the development of a VM culture.

China has been using VM for about three decades now. During this time, VM has made significant contributions to the development of the Chinese economy that has seen steady growth at around 10 percent annually. The policy change and the adaptation of a market economy was a challenge to the state-owned enterprises where VM had seen wide usage. Many Chinese VM experts rate the challenge faced by these enterprises in applying VM in the new environment and the assistance that VM institutions can provide, as the biggest current issues. The lack of evidence of direct savings created by VM in China has dampened continued Central Government support and thus the development of VM in China. A research project has begun at the Department of Building and Real Estate of The Hong Kong Polytechnic University, in collaboration with a number of key universities in Mainland China, to look at these problems and find effective measures to overcome them.

5. Better education of industrial practitioners and owners in VM usage

A 2009 survey among construction industry engineers showed a disturbing tendency. Of the 85 completed surveys, 24 percent of the design engineers and 35 percent of the construction site engineers had 'a total lack of understanding of VM' (Li, 2012). To increase awareness and usage of VM, more effort should be put into educating potential practitioners of VM as well as the owners of companies and projects in which VM can be applied, of the benefits of VM.

An increased understanding of the cradle to cradle cycle of a project will help deliver VM with a wider focus and thus a higher impact. If projects are no longer only considered in separate phases, early VM studies can help increase savings and appeal to a broader audience, especially in the construction industry. To help with the VM education a larger effort on knowledge exchange with foreign countries and VM institutions could greatly benefit VM in China.

6.5   VM Development in Different Countries

VM was developed at the General Electric Company in the USA in 1947 and it has further evolved and spread widely into the engineering and manufacturing sectors in North America. VM has spread throughout the world and is now well received in Europe, Japan, Australia, Korea, India, Saudi Arabia and China. A summary of VM development in Europe is given in Table 6.11.

6.6   VA/VE/VM Standards Worldwide

A summary of Existing VA/VE/VM Standards worldwide is given in Table 6.12.

Table 6.11 VM in Major European Countries

	UK	Germany	France	Italy
Society Name and Acronym	Institute of Value Management (IVM)	Verein Deutscher Ingenieure (VDI) – Zentrum Wert Analyse (ZWA)	Association Francaise pour L'analyse de la Valeur (AFAV)	Associazione Italiana per L'analisi Del Valore (AIAV)
Year of Creation	1966	1974	1978	1985
Legal Status	Non-profit-making organisation	Technical division of the German Association of Engineers	Non-profit-sharing association	Non-profit-sharing association
No. of Members	>160	>600	>800	>230
Regular Publications	Value(Quarterly)	WA-Kurier (Quarterly)	Valeur (Quarterly) Le bulletin de l'AFAV (Monthly)	Valore
Main Activities	Executive meetings, VM awareness seminars	VM Training, Seminars, Conferences	Regional Meetings, Seminars	Developing and spreading the methodology nationally & abroad
National Meetings	No National Conference	Two VM conferences each year	International VM Conference every two years	International VM conference every two years
VM fees/year	9.5 M ECU	18 M ECU	15 M ECU	3.2 M ECU
Certification	Yes (3 Categories.)	Yes (4 Categories.)	Yes (4 Categories.)	None
Standards	None. IVM is moving towards standardization	DIN 69 910: The Value Analysis System – description and work plan	Afnor x 50-150, 151, 152. 153: Definitions of VA; Functional expression; Basic features; Recommendations	None. AIAV is moving towards standardization

Table 6.12 Existing VA/VE/VM Standards Worldwide

Country	Existing Standard
Germany	DIN 69910 'Value analysis'
France	Afnor X 50-150 'Value analysis vocabulary' Afnor X 50-151 'Functional performance specification' Afnor X 50-152 'Basic characteristics of value analysis' Afnor X 50-153 'Recommendations for the implementation of VA'
Australia & New Zealand	AS/NZS 4183:1994 'Value Management'
European	The European Standards Committee (CEN) and the European Commission 'SPRINT' Working Group are jointly working on a VM Standard. The European Standard EN 12973:2000
Austria	Onorm A 6750 – VA: terminology, principles, influences, behaviour Onorm A 6750 – VA: VA and associated activities Onnorm A 6750 – VA: VA unit, organizational status, job description Onnorm A 6750 – VA coordinator: tasks and requirements Onnorm A 6750 – VA modules: principles, terminology, behaviour, work
Hungary	MI8871/T00 – Value analysis: terminology and procedures
Poland	PN81/39100 – Value analysis: terminology
China	GB8223-87 'Value Engineering – Definitions and Procedures'
USA	Value Methodology Standard: 1998
Others	India

6.7 VM Practice in Hong Kong, Australia/ UK, and USA

In Hong Kong, it is observed that VM practice is different from VM practices in both the USA and Australia/UK. A mixed VM, similar to the model used in Australia and the UK, is widely used in Hong Kong's construction industry. A comparison of VM practice in Hong Kong, Australia/UK, and USA is given in Table 6.13.

The duration of VM studies in Hong Kong is 4-16 hours, compared with Australia's 8–24. A one-day (8 hour) workshop is the most popular practice in Hong Kong because clients are often trying to save time and reduce project consultancy fees. Moreover, the size of the study team is larger in Hong Kong because, in addition to members of the project team, many stakeholders are invited from various departments or organisations in order to widen the field of expertise involved in the study. These stakeholders usually constitute a significant proportion of a study team, with their number as high as 70 on a single project (Pickles, 2000).

In summary, the success of VM in improving procurement systems has gained acceptance in Hong Kong's construction industry, and it is becoming ever more important and popular.

Table 6.13 Comparison of VM Practice in Hong Kong, Australia/ UK, and USA

	Hong Kong	*Australia/UK*	*USA*
Stage of Application	Feasibility or concept design	Feasibility or concept design	Sketch or detail design stage
Duration of Study	4–16 hours	8–24 hours	40 hours
Study team	Original project team + Other Stakeholders	Original project team + Other Stakeholders	Independent team
Number of Participants	20–70	15–30	5–8
Job Plan	Information Analysis Creativity Evaluation Development Presentation	Information Analysis Creativity Evaluation Development Presentation	Information Creativity Evaluation Development Presentation
Function Analysis	Not Essential	Not Essential	Essential
Facilitation	Essential	Essential	Not essential
Use FAST	No	No	Yes
Use Function Cost Analysis	Rarely	Rarely	Yes
Target Cost	Rarely	Rarely	Yes

6.8   VM Professional Bodies and Training

VM Certification procedures exist in the USA, UK, Germany, France, Australia, New Zealand and Hong Kong. Other countries such as Japan, India and Korea adopt the SAVE certification procedures. A summary of the existing certification procedures is given in Table 6.14 below. For details of certification procedures in Hong Kong, please refer to the HKIVM website and their facilitators section.

Table 6.14 Existing Certification Procedures

Countries	*Professional Bodies*	*Certification Procedures*
USA	SAVE International	Certified Value Specialist (CVS) Associate Value Specialist (AVS) Value Technician (VT)
UK	The Institute of Value Management	Certificated Value Analyst (CVA) – UK only Professional in Value Management (PVM) – (Europe) Trainer in Value Management (TVM) – (Europe)
Australia	Institute of Value Management – Australia	Fellow => Practitioner Fellow Member => Practitioner Member Affiliate Member
Hong Kong	The Hong Kong Institute of Value Management	Member => Value Management Facilitator (VMF)

6.9   References

Abidin, N.Z. and Pasquire, C.L. (2007). Revolutionize value management: a mode towards sustainability, *International Journal of Project Management*, 25(3), 275–282.

Barton, R. (2002). Integrating values, paper presented at the International *Conference of the Institute of Value Management – Balancing the Scorecard*. Hobart, Australia.

Barton, R., Jones, D., and Andersen, H. (1999). Incorporating the values of ecologically sustainable development into project definitions using soft value management, *Managing Sustainable Values, Proceedings of the International Conference of the Institute of Value Management*, 6–7 May, Hong Kong.

Barton, R., Jones, D., and Gilbert, D. (2000). Initiating sustainable projects, paper presented at *International Symposium on Shaping the Sustainable Millennium*, Queensland University of Technology, Brisbane, Australia.

China Value Engineering (1990). Applications of VE over the past decade, *China Value Engineering*, 1(1).

Dell'Isola, A.J. (1982). *Value engineering in the construction industry*, 3rd edn. New York: Van Nostrand Reinhold Company Inc.

Fong, S.W. (2003). Value management – going all out for knowledge creation, *The Value Manager*, 9(1) available at: http://hkivm.org/publications/03/TVM2003-1.pdf

Fong, S.W. and Shen, Q.P. (2000). Is the Hong Kong construction industry ready for value management? *International Journal of Project Management*, 18(5), 317–326.

Fong, S.W., Shen, Q.P., Chiu, W.I., and Ho, M.F. (1998). Applications of value management in the construction industry in Hong Kong. Hong Kong Polytechnic University, ISBN 962-367-231-4.

Grosvenor, R.F. (1993). Value engineering: what is it? *Hong Kong Engineer*, 21(6).

Han, R. (1998). The development of value engineering (in Chinese), China value engineering in China over the past 20 years (1978–1998). Beijing: *Coal Industry Press*, 87–92.

Ho, C.W. (1995). Design and build: a developing procurement option in Hong Kong, *Discussion paper series no. S0036, Department of Surveying, University of Hong Kong*.

Hu, S.H. (1999). A review of value engineering development (in Chinese), *Value Engineering*, 18(1), 9–10.

Kelly, J. and Male, S. (1988). *A study of value management and quantity surveying practice*. London: Royal Institution of Chartered Surveyors.

Kelly, J.R. and Male, S.P. (1991). *The practice of value management: enhancing value or cutting cost?* London: Royal Institution of Chartered Surveyors Publications.

Lam, P.T.I., Chan, E.H.W., Chau, C.K., Poon, C.S., and Chun, K.P. (2009). Integrating Green specifications in construction and overcoming barriers in their use, *Journal of Professional Issues in Engineering Education and Practice*, ASCE, 135(4), 142–152.

Lam, P.T.I., Chan, E.H.W., Chau, C.K., Poon, C.S., and Chun, K.P. (2010). Factors affecting the implementation of green specifications in construction, *Journal of Environmental Management*, 91(3), 654–661.

Li, H.J. (1998). Applying VE in drainage engineering (in Chinese), China value engineering in China over the past 20 Years (1978–1998). Beijing: *Coal Industry Press*, 399–401.

Li X.Y. and Ma W.D. (2012). Appraisal of value engineering application to the construction industry in China, *Future Wireless Networks and Information Systems: Volume 2, Springer Science & Business Media*, 303–311.

Liu, G.W. and Shen, Q.P. (2005). Value management in China: current state and future prospect, *Management Decision*, 43(4), 603–610.

Liu, X.J. (1998). Construction technical economics. Beijing: *China Construction Industry Press*, 173–191.

Ma, Q.G. (2000). The prospect of VE in China, *Value Engineering*, 19(2), 12–15.

Phillips, M.R. (1999). Towards sustainability and consensus through value management: case study, managing sustainable values, *Proceedings of the International Conference of the Institute of Value Management*, 6–7 May, Hong Kong.

Pickles, L. (2000). Save Tai O, In: *Proceedings of the 4th HKIVM International Conference*, 37–41, November, Hong Kong.

Planning, Environment & Lands Bureau (1998). *Technical Circular No. 9/98, Works Bureau Technical Circular No. 16/98*, Implementation of value management.

Schneider, M. (1999). Value management and sustainability: an opportunity to revolutionalize the construction industry, *Managing Sustainable Values, Proceedings of the International Conference of the Institute of Value Management*, 6–7 May, Hong Kong.

Shen, Q.P. (1997a). Value management in Hong Kong's construction industry: a longitudinal study of clients' perspectives, *2nd International VM Conference, HKIVM*, 12–13 November, Pacific Place Conference Centre, Hong Kong, 1–6.

Shen, Q.P. (1997b). A critical review of VM applications in the construction industry in Hong Kong, *30th Annual Conference of the Society of Japanese Value Engineering*, 17–19 November, Arcadia Ichigaya, Tokyo, Japan, 221–226.

Shen, Q.P. (1997). Application of value management in mainland China: recent developments and future prospects, *Value World, SAVE International*. Dayton, OH.

Shen Q.P. and Liu G.W., (2004). Applications of value management in the construction industry in China, *Engineering, Construction and Architectural Management*, 11(1), 9–19.

Song, Q.R. (1994). Value engineering, *China People Press*, Beijing.

Tang, H. B. and Yang, M. (1996), New value engineering (in Chinese), *Jinan University Press*, 106–111.

Wilson, A.R. (1996). Value management – the next management fad, *International Conference on Value Management in the Pacific Rim*, Pacific Place Conference Centre, Hong Kong, May 29–30.

Xiao, X.D. (1998). A study of the current situation of value engineering in China (in Chinese), China value engineering in China over the past 20 years (1978–1998). Beijing: *Coal Industry Press*, 93–95.

Yeomans, P. (1999). Value management saving the planet: potent paragon or pompous pipedream? *Managing Sustainable Values, Proceedings of the International Conference of the Institute of Value Management*, 6–7 May, Hong Kong.

Zhu, Y.G. (1988). Applications and Development of VE in Shanghai, *Shanghai Value Engineering*, 1(1).

Zou, L.Z. (1998). Applying VE for optimising the design of a bridge (in Chinese), China value engineering in China over the past 20 Years (1978–1998). Beijing: *Coal Industry Press*, 402–405.

7   GDSS in VM Studies

Geoffrey Q.P. Shen and Timmy S.C. Fan

7.1   Introduction

This chapter introduces the features and functions of a group decision support system (GDSS) that has been developed by research. A validation of the proposed system is described and the results are discussed.

7.2   Learning Objectives

Upon completion of this chapter, you should be able to:

1. Explain what is a GDSS
2. Identify the features of GDSS
3. Describe the functions of GDSS
4. Discuss the potential application of GDSS in VM studies

7.3   Introduction

Value Management (VM), also referred to as value engineering and value analysis, is a function-oriented systematic team approach to providing value in a product, system or service (SAVE International, 1998). The implementation of VM in a construction project is normally in the form of one of more workshops attended by major stakeholders and facilitated by a value specialist. The workshops follow a systematic job plan.

As a direct result of technological developments, uncertain economic conditions, social pressures, and fierce competition, construction industry clients have increased their demands on the industry in terms of project quality, costs of delivery, total project time, and, above all, value for money. To help cope with these demands, VM has been widely used in many developed countries for decades.

Reluctance to use VM often comes from the large investment in the expensive team required to undertake the VM process. If the VM process could be made more effective and efficient, the cost of performing VM would decrease. A difficulty faced by VM is the dominance and conformance pressure that can happen when an employer-employee and superior-subordinate situation exist in the same team.

Table 7.1 shows the results of a survey by Shen *et al.* (2004) that revealed the difficulties encountered by VM in the construction industry.

A group decision support system (GDSS) could be used to improve the efficiency and effectiveness of the VM process by utilising the latest technological developments and thus increase the process gains and limit the process losses (Shen *et al.*, 2004). As a type of information technology, GDSS can promote active participation, encourage interaction and facilitate decision analysis among members in the VM process. GDSS is designed to supply discussion support, information support, collaboration support, and decision analysis support for the VM process (Shen *et al.*, 2004). Integrating GDSS with the strength of the traditional face-to-face (FTF) mode exploits the benefits of both modes of communication.

Through research and by integrating the aforementioned modes, the authors of this chapter developed a web-based prototype GDSS to support the VM process and named it the Interactive Value Management System (IVMS). This chapter illustrates the features and functions of the IVMS, describes how the system was validated, discusses results, and presents a conclusion.

7.4 Design of the IVMS

Using the features of a GDSS, the IVMS was designed as shown in Figure 7.1.

Purpose of the IVMS

The IVMS is a useful toolbox to support VM practitioners. It helps overcome or lessen the problems usually associated with traditional VM processes. The IVMS is designed to provide complementary technical support for the traditional VM process, and integration with the FTF method can exploit both modes of communication to the fullest. Computerized records of different types of projects for future

Table 7.1 Problems of VM in the Construction Industry (Shen *et al.*, 2004)

Problem	Reasons	Impact
Lack of information	• Poorly organized project information in the pre-study phase • Difficulty of retrieving project information in meetings	Increases 'uncertainty' in the outputs of VM studies
Lack of participation and interaction	• Shy about speaking in public • Pressure to conform • Dominance by a few individuals • Poor team spirit	Member's contributions are reduced
Difficulty in conducting evaluation and analysis	• Insufficient time to compare analysis • Insufficient information to support analysis	Members have difficulties to complete tasks on time

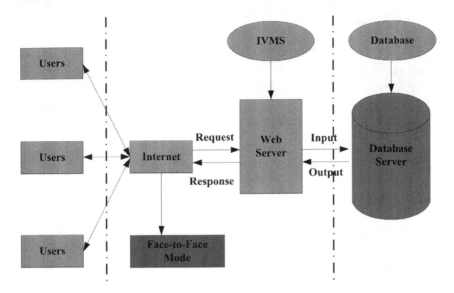

Figure 7.1 IVMS Design

reference can be created with the IVMS, and the system can also be used as a teaching tool to introduce new users to VM.

Development Environment

Microsoft Windows Server 2003 is used as the operating system and Microsoft Access 2002, which supports data access, is used as the database management system. Microsoft Visual Studio .Net 2003 is used as the development environment of the application system, and the system is coded using Active Server Pages.NET, C#, and JavaScript.

Main Features

IVMS is a web-based user-friendly system. The system is installed and operated on a web server, which means that no download is required to run the system and it can be accessed from any computer at any time, so members can get the support for all phases of the VM workshops (pre-workshop, workshop, and post-workshop), from any location.

According the Value Methodology Standard (SAVE International, 1998), a VM process comprises a pre-workshop phase, a workshop phase and a post-workshop phase. The workshops are further divided into phases of information, analysis, creativity, evaluation, development, and presentation. In the following section, the support provided by the system in each phase of the VM process is introduced.

7.5 Proposed Usage of IVMS

A. Pre-workshop Phase

The pre-workshop phase provides an opportunity for all parties to understand project issues and constraints before the VM workshops begin. According to SAVE International (1998), the preparation tasks involve the following:

- Collecting user/customer attitudes
- Completing the data file
- Determining evaluation factors
- Defining the scope of the study
- Building data models
- Determining team composition

As indicated in Table 7.1, one of the main problems in the VM process is lack of information, which is in line with the findings of Park (1993). This is manifest at the pre-workshop phase by poorly coordinated project information. In order to overcome this problem, the IVMS provides an 'Information Centre' for users to store and exchange project information. This can be used in the pre-workshop phase for two major purposes.

Firstly, it can be used as 'My Store' where users can upload information to the system that others cannot access without their permission. My Store can also help users to classify their uploaded information and generally simplify the information storage process.

Secondly, as shown in Figure 7.2, users can share their uploaded information with others. Users who also share information can preview or download the information that others have shared. When new information becomes available, the 'Email Notification' function will notify all users by sending an email to each one. This means that users will always receive the most up-to-date project information. Traditionally, information is exchanged by post or email among the users, which is time-consuming and bears the risk of being lost. In this system, all of the information is securely stored in the Information Centre where users can have instant access to the latest information. Moreover, as the information is always available in the system, having not received it can never be an excuse for inaction.

Through the above functions, the system not only simplifies and shortens the information exchange process but also improves the consistency of information by sharing project information through the Internet. A bulletin board is also provided by the system for users to disseminate ideas and conduct discussions. It can also be used to collect users' opinions. One or more pre-meetings may be held in this phase to allow everyone in the project to understand all the issues and constraints. When the users have some questions or fresh ideas after or before the pre-meetings, they can post their views in the bulletin board where others can read it and reply. It provides users with another way to communicate and exchange information.

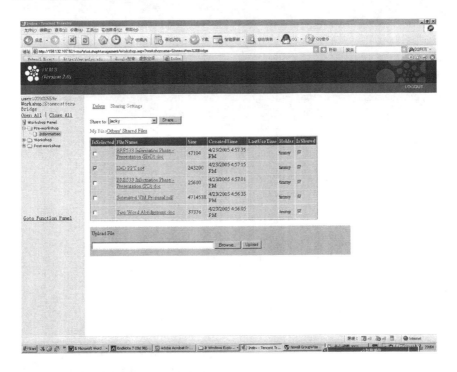

Figure 7.2 Information Centre

Since another purpose of this system is to provide a historical database for VM studies, the information hierarchy has been designed to make it easy to store the data of each VM workshop in the database, as shown in Figure 7.3.

Participant information provides data on who attends VM workshops, while project information contains the project's purpose, assumptions, and other necessary information. The workshop information contains the output from the VM process. Other information, such as information contained in the Information Centre and posted on the Bulletin Board, is also stored in the database and essentially defines the particular VM process by distinguishing it from others. All of the VM workshops conducted using the IVMS are stored in the database for future reference. All of the problems/needs and the corresponding system features in the pre-workshop phase are listed in Table 7.2.

B. Workshop Phase

1. Information

In this phase, information relating to the project such as costs, quantities, drawings, specifications, manufacturing methods, samples and prototypes are collected together for further use (Kelly *et al.*, 2004). Users need to retrieve project information to confirm the objectives, clarify the assumptions and review the scope of the

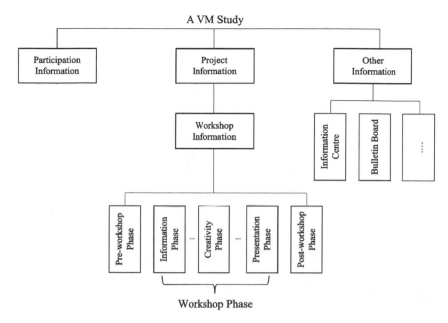

Figure 7.3 Information Hierarchy

VM workshops. However, the difficulty of retrieving project information in VM workshops is problematic.

A conference room where VM workshops are typically held is a semi-closed environment in which the physical boundary may prohibit participants from retrieving new information. By using the IVMS through the Internet, the physical boundaries usually associated with the conference room are broken, and members can easily access external information during the workshop. To increase this capability of the system, an online website database is provided to facilitate information searches. The users may add their own links to the database to enhance it.

The Information Centre is an important component in the process of retrieving information. All project information can be stored in the system prior to the workshop starting; users are then able to easily find the information they need. Once the workshop starts, information that needs to be viewed by all participants can be shown via a projector or on a large LCD screen, or it can be broadcasted through the system to the personal terminals of workshop participants. Both these options make the data retrieval process more efficient.

Table 7.2 IVMS Features and the Problems Addressed in Pre-workshop Phase

System Features	Problems Addressed
Information Centre	Poorly coordinated project information
Bulletin Board	Collecting user/client attitudes
Information sharing & 'My Store'	Information exchange & store

Through the above functions, the IVMS can provide valuable support at every phase of the VM process by vastly improving the availability of information for all users of the system.

2. Analysis

This is the function analysis phase where the project requirements and the required work is clearly defined (Assaf *et al.*, 2000). Invented by Charles W. Bytheway in 1964, the Functional Analysis System Technique (FAST) is one of the main tools that can assist in the function analysis. This technique usually starts off with a brainstorming session, aiming to clarify the functions required by the product or service. 'All functions are expressed as an active verb plus a descriptive noun' (Kelly *et al.*, 2004). The functions that are found are then sorted by the VM team into a diagram. The diagram shows the logical relationship between the functions by placing higher level functions on the left hand side, while lower level functions are placed on the right side, creating a systematic representation of the functions. At the end of the analysis phase, the functions where improvement can be made are selected for detailed study in the next phases of the VM process.

The IVMS helps users during the brainstorming sessions by enabling them to log in to the system's Virtual Meeting Rooms. Depending on the situation, access to these virtual meeting rooms can be switched between anonymous and nominal mode. Regardless of mode, users are able to see all the functions that have been created by all workshop participants. All the functions that are generated are stored automatically and simultaneous in the system, which compared to the traditional way of recording them on paper saves a lot of time. Because the brainstorming is an essential component of the 'creativity' phase, the Virtual Meeting Rooms are detailed in the next section.

After the functions have been prepared in the brainstorming session, the VM team arranges the functions according to 'needs' and 'wants', with the highest 'need' going to the far left side, and the lowest 'want' at the far right side. Commonly used software, like Microsoft Excel and Microsoft Access can be integrated with the system to provide more model tools, and the resulting data analysis process can be displayed with an LCD projector to all workshop participants.

3. Creativity

The purpose of this phase is to create alternative ways to accomplish basic functions as required by the client. To do this, creativity stimulating techniques like brainstorming, synectics, morphological charting, and lateral thinking are used (Shen and Shen, 1999). Of these techniques, brainstorming is by far the most popular used in the creativity phase of workshops. The essence of the approach is for the workshop participants to consider the functions and then contribute with any suggestion that either expands, clarify, or answers the function. However, as shown in the survey referred to earlier in this chapter, some workshop participants are reluctant to contribute in this phase, because they are not comfortable speaking publicly, or because they fear being ridiculed for their suggestions. Further, this part of the workshop can be dominated by a few extrovert individuals, who are not afraid of criticism from others. In these situations, creativity can be reduced.

To help with communication problems, the system has a Virtual Meeting Room feature, which is similar to the 'chat rooms' that are popular throughout the Internet. As one of the basic rules of brainstorming is to have relatively small groups, with no more than eight members (Norton and McElligott, 1995), and VM workshops usually involve 20-30 stakeholders, five virtual meeting rooms are pre-set in the system. If there are too many members in one group, the facilitator should divide them into several smaller units. Each group should have a designated room to conduct their Creativity session in. The participants go to the room they have been assigned, and submit their ideas under the special functions that have been chosen in the Analysis phase. As shown in Figure 7.4, the functions are at the top of the interface to increase their visibility to the members. As an example, the function in this figure is 'Attract Sighting'. This function is set and changed by the facilitator. The left side shows the names of the members in this room. Each participant can then see all the ideas generated by the other members on his own screen. This is good because the ideas of one member might spark related ideas from fellow members in the room, which can turn into a positive spiral of idea creation. This improves the productivity of ideas and the output of the brainstorming session. All of the ideas are stored in the database as they are generated by the members.

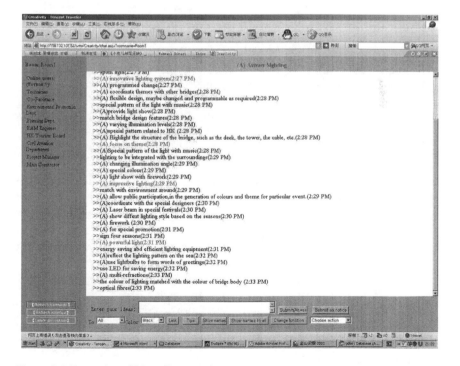

Figure 7.4 Examples of Ideas Generated

These virtual rooms include many features designed to make the brainstorming session more effective and efficient, including:

- **Flexible selectable anonymous or nominal modes** – The public sharing environment can be set to be totally anonymous or nominal, according to the situation. When the environment is anonymous, each user can see on their screen the ideas generated by other group members without knowing who entered them into the system. Participants, who are scared of receiving negative feedback from others in the face-to-face session, may feel more confident in the environment of anonymity provided by the IVMS. This form of anonymity is thought to reduce evaluation apprehension losses (Connolly *et al.*, 1990; Gallupe *et al.*, 1991; Gallupe *et al.*, 1992). 'However, it does not mean that the nominal environment should not be used. While an anonymous environment encourages participants to express their ideas freely, it may also lead to laziness. Some may work hard and some may do nothing or 'free-ride' on the efforts of others. While in a nominal environment, users' names are displayed along with the ideas they generate, giving them the stimulus to generate more ideas to prove themselves' (Fan and Shen, 2011). Hence, the system provides the flexibility for each to choose the discussion mode that best suits them.

- **Parallelism** – 'Parallelism helps to reduce production blocking since users no longer have to wait for others to express their ideas' (Gallupe *et al.*, 1991; Jessup *et al.*, 1990). Users can express their ideas as soon as possible and then go on to generate other ideas.

- **Brainstorming agent** – GDSS-supported groups have a more task-focused communication with less joking and laughing (Turoff and Hiltz, 1982), this leads to people being more critical of each other's ideas when their communication is electronic (Siegel *et al.*, 1986). DeSanctis and Gallupe (1987) also suggested features that are intended to address the social needs of groups to be included in GDSS systems. The IVMS has a built in computer-agent that can pop-up in different situations. If the participation drops, and there is not a lot of activity, the pop-up agent will appear automatically to encourage the participants to engage. The agent can also give the participants some positive feedback when they generate ideas, as shown in Figure 7.5. The system can monitor both the performance of the whole group and of individuals. When an individual participant is silent or active for a while, the agent will pop-up to criticize or praise respectively. As one of main duties of the facilitator during a VM workshop is using his facilitation skills to tap into the group's reservoir of knowledge, experience and creativity, a user who is silent and reserved, may feel embarrassed if approached or advised by the facilitator. The IVMS gives an alternative way of handling such criticism; when participants are criticized by a friendly virtual agent, they may not feel so uncomfortable. In the following validation study, most of the participants found this function useful for improving the atmosphere of creativity.

- **The control functions for the facilitator** – The system control functions are only available to the facilitator of the VM process, these include: changing the environment mode, editing/deleting unnecessary ideas, and posting VM notices. This setting makes it easier for the facilitator to control and conduct the brainstorming session. For example, if someone in the group posts something that the facilitator considers inappropriate or harmful to the brainstorming session, the facilitator could make the workshop go more smoothly by notifying all users accordingly or by expressing misgivings to the individual concerned through the agent.

- **Tips** – This function is designed to inspire the users by providing some relevant ideas, as shown in Figure 7.5. The relevant ideas are generated by the system randomly according to the historical data. Hence, the ideas sometimes can be quite helpful and sometimes can be totally meaningless. However, the purpose of this function is to give some tips, so it should be considered to be successful even if only one idea gives the users some illumination.

- **Other facilities** – Users can set different colours associated with their ideas to make them more attractive and easier to distinguish from others. Internet links can also be posted during the brainstorming process and opened directly through the system.

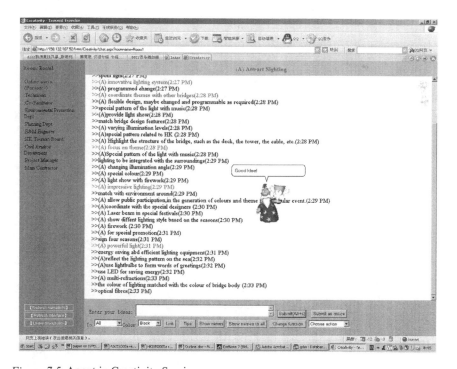

Figure 7.5 Agent in Creativity Session

• **Evaluation** – According to the Value Methodology Standard (SAVE International, 1998), the main tasks of this phase are setting up a number of criteria, and then evaluating and selecting the alternatives that were generated during the creativity phase. Various models and techniques such as, cost models, energy models, LCC models, and the weighted evaluation technique (WET), can be used for this purpose. In addition, some form of weighted vote is also often used (Kelly *et al.*, 2004). This system provides two ways to conduct the evaluation, including the electronic WET and the electronic weighted vote method.

WEIGHTED EVALUATION TECHNIQUE

The information flow of this technique provided by IVMS is shown in Figure 7.6.

• **Idea categorization** – Ideas from the 'Creativity' session will be collected and listed at the corresponding functions automatically. The facilitator can then edit

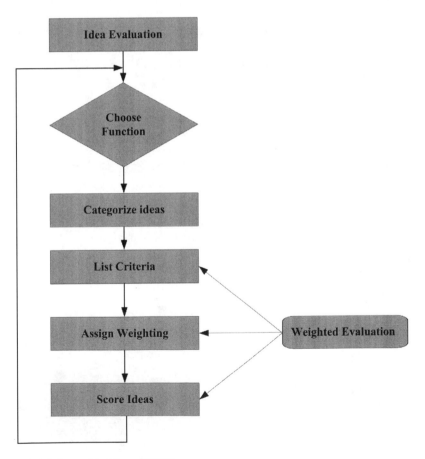

Figure 7.6 Information Flow of WET

and delete any overlapping ideas and correct any the grammar or spelling mistakes that might be present. The ideas are then categorized into P1, P2 and P3 representing ideas that are 'realistically possible', 'remotely possible' and 'fantasy' respectively, as shown in Figure 7.7. Only P1 ideas are considered in the subsequent phases.

- **Assign weighting** – Because the importance of each criterion is different, a relative importance needs to be established and assigned to each of the criteria to help determine how important they are in relation to the other criteria. The system can do this electronically by using a pair-wise method to determine the weighting given to each criterion. Pair-wise is a team-oriented technique to differentiate where a decision has to be made on which of two criteria is the most important. The 'Weighting' screen is designed with this method in mind. The system assigns each criterion with a letter from the English alphabet, and then the preferred criteria are selected from a pop-up list with two fixed entries at a time.

    The first part consists of the letters of the two criteria that are being compared. If criterion A and B are being compared, the entries of the first part will be A and B. The other part is four fixed entries, with 1 being 'Slight, no preference', 2 being 'Minor preference', 3 being 'Medium preference' and 4 being 'Major preference'. Thus, a complete score when comparing criterion A and B, if A is much more important than B, should be A/4. This technique

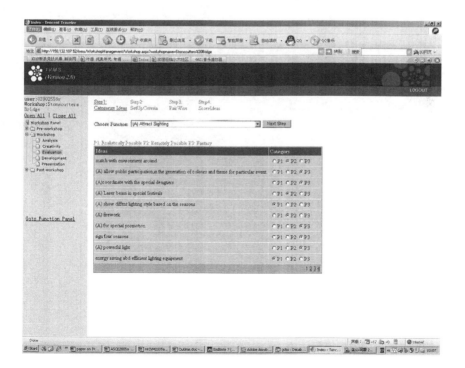

Figure 7.7 Idea Categorization

will be repeated between all pairs of criteria. The final score for A is then the sum of all the numbers associated with A (A/x), in its comparisons with the other criteria. The system will automatically sum this up and show the scores.

- **Score ideas** – An interactive form will be created consistent with the quantity of criteria and the P1 ideas from the previous phases. Since it is better for participants to focus on one criterion only when scoring ideas, the other criteria are blind to participants when they score the ideas according to a criterion. When a user wants to score ideas according to one criterion, he/she should choose the criterion first. Participants must give a score to every idea under each criterion. Each score is then multiplied by the criterion's individual weighting and the resulting total will be automatically computerized as the final score for each idea after the participants have finished the session. When a paired comparison has been conducted, the results are used to score the ideas. Figure 7.8 shows the 'Score ideas' and Figure 7.9 displays the outcome. The results can be revised when there is a divergence among the participants.

- **Weighted Vote Method** – This is another way for users to choose valuable and realistic ideas. Each user needs to select a number of ideas that they think are acceptable, and the ideas which are the more frequently selected will be further considered. At the outset, the facilitator should set a number of ideas that users

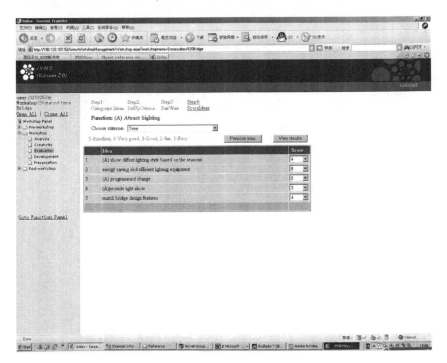

Figure 7.8 Evaluation of Ideas

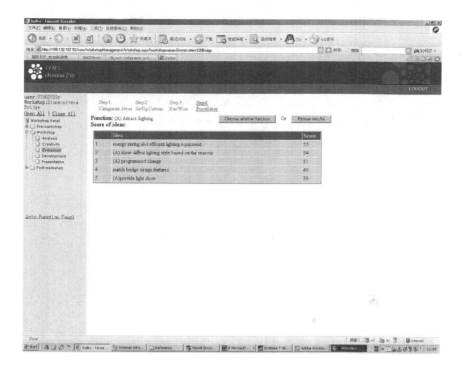

Figure 7.9 Results of Idea Evaluation

could select, which should be less than the total number of ideas. Each user can then vote, and the results of the weighted voting will be calculated automatically. Bar charts of the results are produced in order to make the outcome easy to read. Figure 7.10 shows the interface for voting and the results.

4. Development and Presentation

The development phase examines the selected alternatives in sufficient depth and results in written recommendations for implementation. This involves both detailed technical and economic evaluation and also considerations of the applicability for implementation (Shen and Shen, 1999). There is wide scope for the use of life cycle assessment, cost models, and computer aided calculations at this stage (Kelly *et al.*, 2004).

The presentation phase defines and quantifies results to prepare and present a Value Management Change Proposal (VMCP) to the final decision makers. The IVMS information centre makes this presentation more flexible and effective. Users can upload the proposal to the information centre, and then others are able to read it through the system. The system will list all the outcomes of the workshop for users to download or preview on line. All of these features will help users to save time and communicate workshop outcomes quickly.

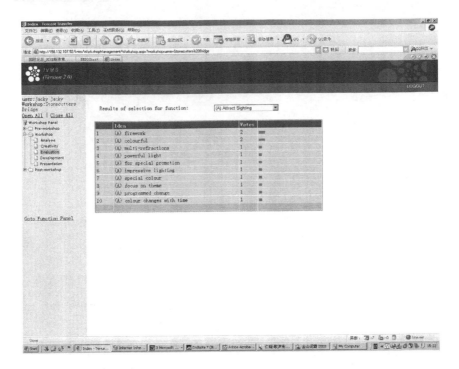

Figure 7.10 Weighted Vote Results

C. Post-workshop Phase

The objective of this phase is to assure the proper implementation of the approved change recommendations. Assignments are carried out to track the progress and collect feedback on the proposal (SAVE International, 1998).

The system features an automatic collection process that collects all the information from the workshop, including the number of ideas generated and the timing of each phase. This information can be used to evaluate the process and outcomes of the workshop. To review the workshop, a questionnaire is presented to the participants at the end of the workshop to collect their views on the processes and the outcomes of the workshop. The information and the questionnaire are used to produce a score for the workshop. While this score does not indicate how the VM process improved or benefited the project, it does give the participants a general picture of how successful the workshops were perceived to be.

The system also has several other ways to collect feedback, such as a bulletin board, information centre, and notice board. Through the use of these features the users can conduct online discussions, post their ideas, and submit their feedback on the workshop. The flow of VM workshops using IVMS is shown in Figure 7.11.

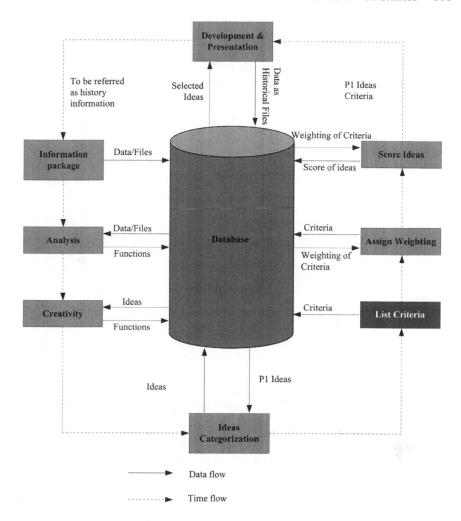

Figure 7.11 Flow of VM Workshops Using IVMS

7.6 Validation of IVMS

When trying to build a computerized conferencing system, Hiltz and Turoff (1981) found one of the main issues to be that users could not accurately predict what they needed prior to using the system. Consequently, users must have extensive experience with GDSS before the effectiveness of the design of the system can be assessed (DeSanctis and Gallupe, 1987). In order to investigate the effectiveness and efficiency of the IVMS, a VM study was conducted on the system in April 2005.

The participants of the study were made up of twenty part-time research students from the Department of Building and Real Estate, The Hong Kong

Polytechnic University. All of the students had several years of work experience from the construction industry. For this VM study they represented the stakeholders relevant to the task, including client, designer, quantity surveyor, and government representatives. Because of their work experience, many of the participants played roles they were familiar with from working on real life projects. Their experience meant that they would think similarly to participants in a real-life VM workshop.

The case used for the study was a real-life project from Hong Kong, rather than a made-up one. The participants were asked to organize a GDSS-based VM workshop process, including the pre-workshop, workshop, and post-workshop phases. All the participants of the study were required to attend a training session on the usage of IVMS in VM workshops, to ensure their familiarity with its use.

A GDSS room, shown in Figure 7.12, was created for the study. Each of the participants was given a laptop and a wireless network was set up in the room so each user could access the system from their provided laptop throughout the workshop. A projector and a large screen were also available in the GDSS room, where notices and group information could be displayed for all to see.

To understand the nature and strength of support that the IVMS can provide, a questionnaire survey was conducted during this study. The questionnaire was based on research conducted by Shen *et al.* (2004). Each item was measured on a five-point Likert-type scale, with 1 as 'strongly disagree' and 5 as 'strongly agree'. The survey results indicate that the use of IVMS in a VM workshop provides valuable

Figure 7.12 A VM Study with GDSS Support

support to the team in many ways. Table 7.3 shows that most of the IVMS functions are found to be very useful in supporting and improving the VM process; scores for 9 out of the 12 items are above 4.00, while 3 of the remaining items are above 3.00. This is strong evidence that GDSS tools can be used to improve the VM process. The results also show that the questionnaire statements 'IVMS can promote active participation in idea generation' and 'IVMS can avoid conformance pressure in idea generation' were the highest rated answers by the participants. As mentioned earlier, pressure to conform and a lack of active participation are two of the main problems faced in VM workshops; use of the IVMS can help mitigate them.

Ranked as the second most useful function, the feature that allows the IVMS to improve the availability of information and enhance the information exchange process is important as this can help improve the lack of information in the VM process. There is also agreement from the survey participants that the IVMS is useful in the analysis and evaluation phases. The interface of the IVMS is very important for its usefulness in the VM process; if workshop participants do not feel comfortable with the interface, performance of the whole VM process will be negatively affected. A score of 4 for 'I feel comfortable with the current interface of the IVMS', as shown in Table 7.3, indicates that most of the participants agree with the statement.

The results of this study suggest that the support provided by the IVMS can alleviate or overcome the problems and difficulties involved in the VM process.

Table 7.3 Summary of the Survey Results on the Support of IVMS (Fan *et al.*, 2012)

Type of Support	Average
Support in Information phase	
IVMS can improve the availability of information.	4.18
IVMS can improve the information exchange process.	4.18
Support in Function analysis phase	
IVMS can simplify the function analysis processes.	4.12
IVMS can enhance the function analysis processes.	4.18
Support in Creativity phase	
IVMS can promote active participation in idea generation.	4.29
IVMS can avoid conformance pressure in idea generation.	4.24
IVMS can prevent domination in discussion.	3.94
The pop-up character in IVMS can enhance the atmosphere of creativity.	3.88
The function of 'Tips' can help me in generating ideas.	3.47
Support in Evaluation phase	
IVMS can simplify the evaluation processes.	4.06
IVMS can enhance the evaluation processes.	4.00
Interface of IVMS	
I feel comfortable with the current interface of IVMS	4.00

5: Strongly agree – 4: Agree – 3: Neutral – 2: Disagree – 1: Strongly disagree.

7.7   Conclusions

As a web-based GDSS, the Interactive Value Management System (IVMS) has been developed to help alleviate the problems usually associated with traditional VM workshops. The major functions of the system, typical usage scenarios, and validation of the system are presented in this chapter. The system is a useful addition to VM workshops as a tool for facilitating the information exchange process, and for encouraging greater interaction and more active participation in VM workshops.

7.8   References

Assaf, S., Jannadi, O., and Al-Tamimi, A. (2000). A computerized system for application of value engineering methodology, *Journal of Computing in Civil Engineering*, 14 (3), 206–214.

Briggs, R.O., Balthazard, P.A., and Dennis, A.R. (1996). Graduate business students as surrogates for executives in the evaluation of technology, *Journal of End-User Computing*, 8(4), 11–17.

Connolly, T., Jessup, L. M., and Valacich, J.S. (1990). Effects of anonymity and evaluative tone on idea generation in computer-mediated groups, *Management Science*, 36 (6), 689–703.

DeSanctis, G. and Gallupe, R.B. (1987). A foundation for the study of group decision support systems, *Management Science*, 33(5), 589–609.

Fan, S.C. and Shen, Q.P. (2011). The effect of using group decision support systems in value management studies: an experimental study in Hong Kong, *International Journal of Project Management*, 29(1), 13–25.

Fan, S.C., Shen, Q.P., and Kelly, J. (2008). Using group decision support system to support value management workshops, *Journal of Computing in Civil Engineering*, 22(2), 100–113.

Fan, S.C., Shen, Q.P., and Luo, X.C. (2010). Group decision support systems in value management, *Construction Management and Economics*, 28(8), 827–838.

Fan, S.C., Shen, Q.P., Luo, X.C., and Xue X.L. (2012). A comparative study of traditional and group decision support systems (GDSS) – supported value management workshops, *Journal of Management in Engineering*, 29(4), 345–354.

Fjermestad, J. and Hiltz, S.R. (1999). An assessment of group support systems experimental research: methodology and results, *Journal of Management Information Systems, 15(3)* 7–149.

Gallupe, R.B., Bastianutti, L.M., and Cooper, W.H. (1991). Unblocking brainstorms, *Journal of Applied Psychology*, 76(1), 137–142.

Gallupe, R.B., Dennis, A.R., Cooper, W., Valacich, J.S., Bastianutti, L.M., and Nunamaker, J.F. (1992). Electronic brainstorming and group size, *The Academy of Management Journal*, 35(2), 350–369.

Hiltz, S.R. and Turoff M. (1981). The evolution of user behaviour in a computerized conferencing system, *Communications of the ACM*, 24(11), 739–751.

Lorge, I., Fox, D., Davitz, J., and Brenner, M.A. (1958). A survey of studies contrasting the quality of group performance and individual performance, 1920–1957, *Psychological Bulletin*, 55(6), 337–372.

Jessup, L.M., Connolly, T., and Tansik, D.A. (1990). Toward a theory of automated group work: the de-individuating effects of anonymity, *Small Group Research*, 21(3), 333–348.

Kelly, J., Male, S., and Graham, D. (2004). *Value management of construction projects*. Oxford: Blackwell Science Ltd.

Norton, B.R. and McElligott, W.C. (1995). *Value Management in Construction: A Practical Guide*. Basingstoke: Macmillan Press.

Park, P.E. (1993). Creativity and Value Engineering Teams, *Proceedings of the 28th SAVE International Annual Conference*. Fort Lauderdale, FL: SAVE International.

SAVE International (1998). *Value methodology standard*, 2nd edn. Northbrook: SAVE International, 4–18.

Shen, Q.P, Chung, K.H., Li, H., and Shen, L.Y. (2004). A group system for improving value management studies in construction, *Automation in Construction*, 13(2), 209–224.

Shen, Q.P. and Shen, L.Y. (1999). Value management as a vehicle for scope management of construction projects, *Journal of Harbin University of Civil Engineering and Architecture*, 32(5), 107–115.

Siegel, J.V., Dubrovsky, V., Kiesler, S., and McGuire, T.W. (1986). Group processes in computer-mediated communication, *Organizational Behavior and Human Decision Processes*, 37(2), 157–187.

Turoff, M. and Hiltz, S.R. (1982). Computer support for group versus individual decisions, *IEEE Transactions on Communications*, 30(1), 82–91.

8 Value Briefing

Ann T.W. Yu and John Kelly

8.1 Introduction

This chapter introduces the application of VM to the briefing process known as Value Briefing.

8.2 Learning Objectives

Upon completion of this chapter, you should be able to:

1. Provide a definition of Value Briefing
2. Explain the benefits of using VM for the briefing process
3. Describe the methodology of Value Briefing
4. Apply VM to the briefing process

8.3 Definition of Briefing

The briefing process, also known as architectural programming in the United States, is the first and most important step in the design process, where client requirements for a project are defined and the major commitment of resources is made. It is the procedure of gathering, analysing, and synthesising the information needed to inform decision making and decision implementation at the strategic and project planning stages of the construction process. According to the Construction Industry Board (CIB) 1997, briefing is the process by which a client informs others of specific needs, aspirations and desires, either formally or informally, and a brief (or programme) is a formal document which sets out a client requirements in detail.

The briefing process is both critical to the successful delivery of construction projects and problematic in its effectiveness. Problems in projects can often be traced back to the briefing process. The infamous Pruitt-Igoe housing project in the USA was demolished in 1976 because it did not respond to the behavioural and social needs of the users (Duerk, 1993). This building's design failure illustrates very well that a systematic identification of client requirements is a pre-requisite to project success.

Irrespective of how the briefing is undertaken, it is a team activity. The brief writer engages with a finite team of people in collecting, collating and processing information that leads to the performance specification of the project. The brief writer has a choice of either carrying out an exploration through interviews with all stakeholders, or engaging with people in a workshop or series of workshops. These two methods are referred to here as *Investigative Briefing* or *Facilitative Briefing*.

Investigative Briefing

Investigative Briefing involves compiling the brief through a process of a literature review, interviews and meetings with key client representatives, post-occupancy evaluation, and existing facilities walk-throughs. The data thus gathered is checked through presentations at team meetings.

The advantages of the investigation approach are as follows:

- A skilled brief writer is able to logically collect data using proven techniques in an efficient manner, minimising client representative input.
- It is not necessary to identify all stakeholders at the commencement of the briefing. If an important stakeholder comes to light during the process, he/they can be interviewed in turn.
- It is more likely that, through a confidential interview, particularly on a one to one basis, the truthful views of juniors in a hierarchy can be obtained and/or hidden agendas revealed.
- An interview is likely to reveal the decision makers even if their identity previously may not have been made clear.

The disadvantages of the investigative approach are as follows:

- Points raised in later interviews may require revisiting earlier interviewees to clarify issues. Checks need to sometimes be made to ensure all interviewees are using language and terms in a common manner.
- Where the brief writer is not knowledgeable in terms of the client business it may be necessary to enlist the help of an expert to ensure the right questions are asked.
- It is difficult to counter the 'wish list' syndrome, particularly where a forceful stakeholder puts over their requirements as a fait accompli and validation checks with the briefing team may not be possible until all of the interviews have been collated. Discrepancies may therefore only be revealed late in the briefing process.

Facilitative Briefing

In *Facilitative Briefing,* a facilitator independent of the client and the design team guides the whole team through a process of briefing largely using the techniques of value, risk and project management. It is still necessary to undertake a literature review, conduct pre-workshop meetings, post-occupancy evaluation and walk-throughs of existing facilities as means of providing information to the workshop, the results of which are then supplemented and interpreted by the workshop team.

The workshop team comprises stakeholders appropriate to the stage of the briefing exercise, principally whether it is a Strategic Briefing or Project Briefing exercises. Typically a Strategic Briefing exercise takes between 4 and 8 hours and involves between 6 and 20 stakeholders. A Project Briefing exercise typically takes one to three days and involves between 12 and 20 stakeholders.

Advantages of the facilitation approach are as follows:

- A Facilitated Briefing exercise extracts all the information in the shortest time.
- The team contains all the experts necessary to feed information into the project and asks appropriate questions of others.
- Misunderstandings and/or disagreements between team members can be resolved immediately in the workshop.
- The facilitator or other members of the team can challenge 'wish lists'.
- The facilitator is able to summarise the data contributed by the team at stages during the team exercise, therefore the brief largely comprises a collation of these conclusions.
- An intensive, focused, Facilitated Briefing exercise encourages good team dynamics and effective team building highlighted by some clients as the most important aspect of the process and the one least able to be replicated through the investigation process.

Disadvantages of the facilitation approach are as follows:

- A Facilitated Briefing exercise demands, concurrently, the presence of all stakeholders.
- Hidden agendas may be difficult to uncover in a team exercise.

8.4   Stages of Briefing

There are two major stages of the briefing process: The *Strategic Briefing* and the *Project Briefing*. This two-step approach is apparent in both practice and in briefing guides. This structure is due to the nature of the early stage design problems. First, strategic management identify organisational needs during the Strategic Briefing study and decide whether a project (or projects) of a generic type and in a specific location is the optimal solution to the needs. Afterwards, tactical management make decisions on the performance specifications of the project within the activities that have to be accommodated according to the Project Briefing study.

Strategic Briefing

The primary objective of the Strategic Briefing study is to develop a strategic brief which describes in business language the reason for investment in a physical asset, and the purpose of that investment for the organisation together with its important parameters. Hence, questions to be answered include: 'Why invest?' 'Why invest now?' 'What is the purpose of the investment?'

The mission of the project and its strategic fit with the corporate aims of the client organisation is described clearly and objectively in the Strategic Briefing study. These corporate aims are explicit in terms of commercial objectives and usually implicit in terms of cultural values. The combined commercial objectives and cultural values form the clients value system. The study also includes establishing the outline budget and programme.

The Strategic Briefing study explores a range of options for delivering the project, which might be the creation of a new building, additional and alteration works, or refurbishment of existing building(s). The Strategic Briefing study structures information in a clear and unambiguous way to permit the 'decision to proceed' to be taken in the full knowledge of all the relevant facts at the time.

On completion of the Strategic Briefing study, the decision to proceed can be made with confidence, given that all relevant issues and options have been addressed and explored, and alternatives examined.

Project Briefing

The Project Briefing study is focused on how to deliver the technical aspect of the project. It can be described as the construction industry's way of dealing with the client requirements that are expressed in the Strategic Brief. The Project Brief is a translation of the Strategic Brief into construction terms, going in depth with the performance requirements of each of the elements of the project. The Project Brief also includes the spatial relationships of the project, and is the basis on which the design can be developed.

8.5   Benefits of Value Briefing

The benefits of using VM in the briefing process include:

- Achieving value for money from the whole project
- Attaining appropriate quality for the project
- Systematically identifying client's requirements, clarifying their needs versus wants and prioritising options for their issue
- Openly discussing and clearly identifying project objectives
- Evolution of a design with an agreed framework of project objectives
- Creating more alternative solutions
- Validating design proposals
- Enabling clients to participate fully in the briefing process
- Being responsive to client's priorities
- Gaining a client and user insight into the project
- Providing an opportunity for stakeholders to formally participate in the design process
- Facilitating communication between clients and other stakeholders
- Resolving conflict among stakeholders
- Team working to identify opportunities available for development and to highlight any potential problems from the beginning of the project

8.6   Timing for Value Briefing

The Strategic Briefing should be carried out at a time prior to the decision to proceed is made, and should be completed after the workshop. Anybody reading the Strategic Briefing should be able to understand why an organisation has decided to invest in a physical asset or assets and pursue no other strategic options that might compete for the same investment resource at that time. It is recommended that Strategic Briefing should be fixed before detailed design commences in order to avoid redesign and rework.

Project Briefing should be carried out prior to the completion of the project feasibility study in order to derive the greatest benefits from limited resources. A Project Briefing, which forms the basis of the design, should be completed after the workshop and before scheme design.

Please note that the recommended timings are indicative and the exact timing of Strategic Briefing and Project Briefing for different projects may vary slightly according to the scope and complexity of the projects. If circumstances change, Strategic Briefing or Project Briefing may be required again, but to save time the number of reviews may be reduced if the first briefs are well documented and relevant.

8.7   Stakeholders in Value Briefing

Appointment of facilitator

It is necessary to appoint a VM trained facilitator to lead the VM study and to chair meetings. The facilitator must be proficient and experienced enough to take on this key role and must be able to:

- Manage the whole VM process professionally
- Maintain a climate conducive to participation, listening, understanding, learning, and creating

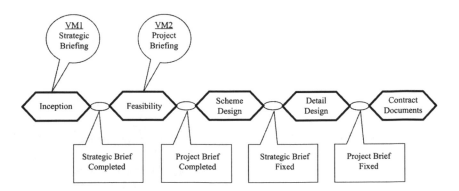

Figure 8.1 Indicative ideal timing of Strategic and Project Briefing

- Listen actively
- Help the team to establish and accomplish its own objectives
- Provide structure and guidance to increase the likelihood that objectives will be accomplished
- Keep the group focused on its objectives
- Encourage relevant dialogue and interaction among participants
- Suggest and direct processes that empowers and mobilises the group to get its work done
- Encourage the group to evaluate its own progress or development
- Capitalise on the differences among the group
- Remain neutral on content and be active in suggesting and directing the process
- Protect group members and their ideas from being attacked or ignored
- Use facilitation skills to tap the group's reservoir of knowledge, experience and creativity
- Sort, organise and summarise group inputs or help the group to do so
- Help the group move to a healthy consensus, define and commit to the next steps, and reach timely closure

The facilitator may or may not be an existing member of the project team. Appointing a facilitator from outside your organisation usually gives the role and the person an air of impartiality and therefore greater credibility among the briefing team members.

Please refer to the following websites for details of qualified facilitators:

- Hong Kong Institute of Value Management (HKIVM) – www.hkivm.org/
- Institute of Value Management UK (IVM) – http://ivm.org.uk/
- SAVE International USA – www.value-eng.org/
- Institute of Value Management Australia – http://ivma.org.au/

The briefing team

The facilitator of choice assists in selecting the briefing team for a specific project, which ideally:

- Aims to be objective
- Has a good understanding of the project's objectives and constraints
- Includes a representative from each of the key disciplines involved
- Focuses on collaboration and trust
- Commits to democratic practice
- Is capable of lateral thinking; team members should be able to think out of the box innovatively
- Includes good team players
- Includes clear communicators

Criteria for an effective briefing team

1. For maximum performance a briefing team should be temporary, formal, task orientated, and has an upper limit of approximately 20 members and a formal leader.
2. A large briefing team (13 to 20 members) requires formal rules and procedures.
3. The briefing exercise should be efficiently conducted over a sufficient but short period of time to prevent the emergence of an informal leader.
4. The briefing team should be project focused and interactive and be comprised of individuals willing to sacrifice individualism for collectivism.
5. Team membership should be effective and balanced as indicated by the ACID Test (see section 8.9).

Potential participants for Strategic Briefing Workshop: 10 to 20, all at senior level in the client organisation.

Potential participants for Project Briefing Workshop: 15 to 20, representatives of the client's managerial team, design and project management team, contractor's team and end users.

8.8   Venue of Value Briefing

Venue

If possible, the ideal venue in which to assemble the participants should be a hotel or resort, which is away from the office. The objective is to enable participants to give their undivided attention to the workshop for a concentrated period of time.

Facilitator

The facilities and equipment needed in a briefing workshop include:

*   Large room with good ventilation and lighting
*   Seating arranged in a U-shape or a semi-circle
*   Overhead projector
*   LCD projector
*   Flip charts and stands
*   Wall space to pin up flip charts
*   White boards
*   Marker pens
*   Post-it notes
*   Blu-tack
*   Sticky dots (red and black)

Additional facilities, such as photocopying, computing, printing, fax and telephones, should be accessible but only when required to provide information for the workshop.

Room Layout

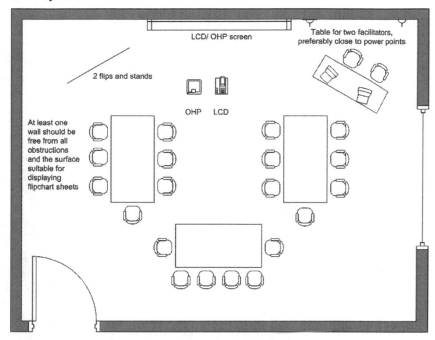

Figure 8.2 Typical room layout of the briefing workshop

Figure 8.3 Photo of the workshop venue

8.9   Tools and Techniques used in Value Briefing

ACID Test – Selecting the briefing team

The team facilitator works either with the client in selecting the members of the briefing team in the pre-workshop phase or can request the client to build the team based simply on the inclusion of those stakeholders with an input relevant to the particular stage in the project development. The ACID Test is to determine precisely who should be a member of the briefing team. The criteria are as follows.

A   AUTHORISE – Include those who are appropriate in connection with the stage of the project development under discussion and who can therefore authorise actions appropriate to any decisions made. Such people have executive authority and are invaluable members of the management team. Their value is in 'their ability to immediately sanction a particular line of discovery or take a decision during the workshop which resolves an issue or unblocks a particular line of investigation' (Kelly *et al.*, 2004)

C   CONSULT – Include experts who have to be consulted regarding particular aspects of the project during workshop discussions. Such people can then be relied upon if a particular line of investigation is dependent upon consultation with an absent expert, thus preventing workshop progress being compromised.

I   INFORM – Do not include those whose major input is simply to be in receipt of decisions reached at the workshop.

D   DO – Include those who are likely to be designated to carry out major tasks specified at the workshop stage. Such players are those, for example, employed to design or construct and need to be fully conversant with the decisions taken during the workshop.

The facilitator and/or project manager may also take into account the following factors when selecting the briefing team.

1.   'Limit multiple representations from one organisation or one department. For example, three members from one organisation where other organisations have single representation will lead to a weight of argument in favour of the multi-represented organisation' (Kelly *et al.*, 2004).
2.   Understand the hierarchical mix (senior, subordinate) within the team.
3.   Understand the relationships between team members, for example, one member may be dependent financially on another team member.
4.   Consider the completeness of the team. Discuss with the client any apparently missing representatives of crucial areas of expertise.

Function Analysis

This section introduces the concept of function analysis and its use at the Strategic Briefing stage of a project to derive the project's mission through function diagramming. The technique of function diagramming is attributed to a VE practitioner, Charles Bytheway, who gave it the acronym FAST (function analysis system technique)

(Kelly *et al.*, 2004)

One of the objectives of the use of function analysis at the Strategic Briefing stage is to lay the foundation for the 'best value for money' solutions to the project problems.

The function analysis technique relies upon the discovery of all relevant information through the issue analysis and the structuring of that information in a way that leads to the recognition of the primary objective of the project.

(Kelly *et al.*, 2004)

There are three steps to the construction of a function diagram:

1) Generation of functions

A function is the specific purpose or intended use for a product, it is the characteristic that makes it work, sell, generate revenue, or meet requirements (Dell'Isola, 1982). The generation of functions starts with the facilitator leading the team to creatively explore the functions required by the project. 'These functions may be high order executive functions or relatively low order 'wants'. All functions are expressed as an active verb plus a descriptive noun, and are recorded on sticky notes and scattered randomly across a large sheet of paper. The facilitator will continually prompt the team to generate functions by referring back to the information from the issue analysis and timelines' (Kelly *et al.*, 2004). A typical list of creatively explored functions is shown in Table 8.1.

2) Sorting of functions / constructing project function priority matrix

After the 'completion of the brainstorming session, the team is invited to sort the notes into a more organised form by putting the highest order "needs" into the top left-hand corner of the paper and the lowest order "wants" into the bottom right-hand corner' (Kelly *et al.*, 2004). The 'needs' are the essential requirements of the project that must be present to serve the client's intention with the project, and the 'wants' are the additions that would be positive to accomplish, but are not necessarily needed. A project function priority matrix is made up from the responses collected from each of the 'sticky notes' as to whether the function is technical or strategic in nature, and whether it is a 'need' or a 'want'. The note is inserted into the appropriate box in the matrix, where it is sorted relative to the other functions in the box (Table 8.2). The sorting is done according to priority, with the highest priorities at the top of their respective boxes. 'It should be emphasised to the team that this procedure is an iterative process and therefore any team member is entitled to move a previously ordered sticky note. Although this activity sounds confrontational it is very rare for disagreement to occur and ultimately the correct ordering of all the functions can be achieved' (Kelly *et al.*, 2004).

Table 8.1 Typical list of creatively explored functions for a residential development

• Secure environment	• Please neighbourhoods
• Establish multi-function integration	• Reduce government accommodation
• Enhance facilities	• Transfer responsibility to private sector
• Ensure user comfort	• Protect existing building
• Meet community needs	• Establish local community pressure
• Create pleasing environment	• Satisfy compensation
• Upgrade living standard	• Enhance environment
• Improve user interface	• Secure funding
• Better utilization of land	• Create value
• Change community perception	• Provide recreational space
• Minimize nuisance to public	• Enhance communication
• Reduce deterioration	• Reduce mosquitoes
• Control finances	• Reduce dust
• Control operational cost	• Preserve parking
• Relieve compliant	• Suppress vibration
• More recreational facilities	• Reduce noise
• Improve accessibility	• Maintain access
• Increase flexibility	• Monitor environment
• Allow private participation	• Monitor dust
• Ensure buildability	• Satisfy safety
• Ensure operability	• Improve safety standard
• Ensure building lifetime	• Limit cost
• Establish project brief	• Control programme

3) Construction of function diagram

A strategic or customer oriented FAST diagram is constructed by focusing on the strategic needs and wants (Figure 8.4). The 'highest order needs tend to form the mission of the project with supporting functions being positioned to the right. The strategic wants tend to be positioned below the centreline of the mission statement' (Kelly *et al.*, 2004). The 'mission statement will require to be word crafted to make it read as a flowing statement. It is important to have unanimous team agreement that the statement truly reflects the mission of the project, remembering that the mission at this stage does not necessarily imply a building' (Kelly *et al.*, 2004).

Table 8.2 Example of a project function priority matrix

Strategic needs	Technical needs
• Secure environment	• Allow private participation
• Establish multi-function integration	• Ensure buildability
• Enhance facilities	• Ensure operability
• Ensure user comfort	• Ensure building lifetime
• Meet community needs	• Establish project brief
• Create pleasing environment	• Control programme
• Upgrade living standard	• Improve safety standard
• Improve user interface	• Limit cost
• Better utilization of land	
• Change community perception	
• Minimize nuisance to public	
• Reduce deterioration	
• Control finances	
• Control operational cost	
• Relieve compliant	
• More recreational facilities	
• Improve accessibility	
• Increase flexibility	
Strategic wants	**Technical wants**
• Please neighbourhoods	• Reduce mosquitoes
• Reduce government accommodation	• Reduce dust
• Transfer responsibility to private sector	• Preserve parking
• Protect existing building	• Suppress vibration
• Establish local community pressure	• Reduce noise
• Satisfy compensation	• Maintain access
• Enhance environment	• Monitor environment
• Secure funding	• Monitor dust
• Create value	• Satisfy safety
• Provide recreational space	
• Enhance communication	

User Flow Analysis

The User Flow Analysis is the first step in the process of deriving a specification of functional space, the foundation of the project brief. This exercise involves the following steps:

Determine users

The first stage is to identify all users of the building. Table 8.3 lists the users of a community office project.

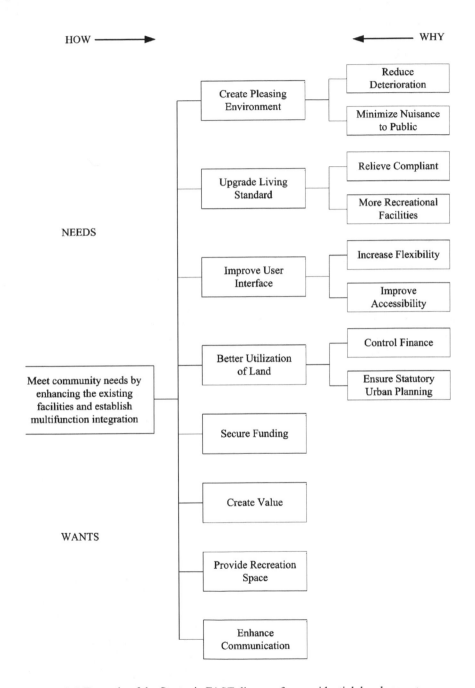

Figure 8.4 Example of the Strategic FAST diagram for a residential development

Table 8.3 List of users for a community office project

- Chief Executive
- Treasurer
- Area Manager
- Housing Administration Staff
- Housing Maintenance Manager
- Housing Maintenance Operatives
- Social Workers
- Youth and Community Workers
- Receptionist
- Filing clerk
- General Public
- Rent Collectors

Kelly *et al.*, 2004.

Flowcharting exercise

Each user from the list in step 1 is studied in turn and a flow chart of their use of space is prepared. This is undertaken by anticipating each activity as part of the user's daily routine within the building where each activity is connected by arrows to the next activity. It is presumed that each activity will require space. Even the activity of entering the building will require an entrance lobby of some sort, and the activity of moving from one space to another indicates circulation space.

(Kelly *et al.*, 2004)

Figure 8.5 and Figure 8.6 illustrate simplistically the activities of the social workers and housing staff in a community office project.

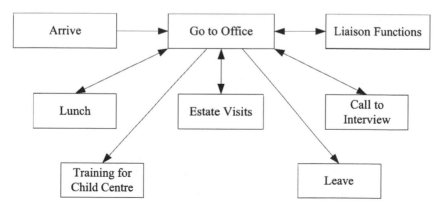

Figure 8.5 Social worker flow chart
Adapted from Kelly *et al.*, 2004.

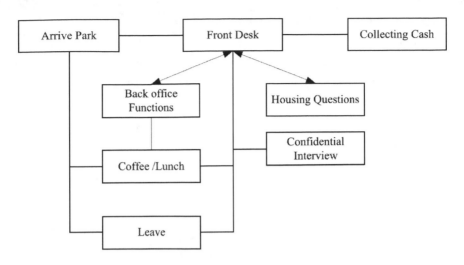

Figure 8.6 Housing staff flow chart
Adapted from Kelly *et al.*, 2004.

Functional Space Analysis

Each functional activity in the user flow diagrams leads to a functional space
used in a particular way.

(Kelly *et al.*, 2004)

For example, the following user spaces can be identified from the results of the
completion of all the user flow diagrams in the user flow analysis.

Room Data Sheets

The project brief should include room data sheet for each functional space.

Each space will have the attributes of size, servicing (heating, lighting, ventilation, and acoustics), quality (normally defined by fittings and furnishings)
and finally the technology support required.

(Kelly *et al.*, 2004)

These room specifications become the raw data from which to compile the room
data sheets. It is recommended that the briefing team should agree the format of the
room data sheets in the workshop while the details of the room data sheets should
be completed in the post-workshop phase. Table 8.5 and Table 8.6 show two
examples of a room data sheet.

Table 8.4 List of functional space identified for a community office project

---

- Car park – secure (council vans, 11 private cars)
- Briefing Area – Rent Collectors, Maintenance
- Cash up Secure Area for Records
- Lunch – Coffee Facility
- Toilets/Showers / Changing / Boot Area
- Maintenance Desk (1)
- Meeting Area
- Reception Area
- Confidential Interview (2)
- Housing Desks (4)
- Storage for Materials/Tools
- Strong Room/Safe
- Pram Park/Cycle Park - secure
- Children's Play Area
- Toilets
- Waiting Area
- Social Worker Desk (1)
- Area Manager Desk (1)
- Youth and Community Desk (1)
- Receptionist/Filing Clerk
- General Office (8)
- Area Manager Office (1)

Adapted from Kelly *et al.*, 2004.

Functional Performance Specification

Functional performance specification (FPS) is defined as a document by which an enquirer expresses a need in terms of user-related functions and constraints (BS EN 12973: 2000). For each of these functions, evaluation criteria are defined together with their levels, with a certain degree of flexibility being assigned (Masson, 2001).

FPS assists the client representatives to:

- Express the needs of the clients and end-users in functional terms, without reference to the technical solutions that may satisfy them, and with a minimum of constraints.
- Stimulate the communication between clients and designers to optimise the design and to find the best proposal.
- Facilitate the examination and comparison of proposals to determine the best proposal for the design.

Table 8.5 Typical room data sheet of a reception area

Room data sheet

Activity Space	Reception	No. 14.2
Occupants	3 receptionists at peak times	
Workstations	3 no.	
	One facing each entrance	
	Nominally wet/dry tills	
	Centre position of enquires	
	Bookings, disabled ticketing, equipment return	
Fittings	3 LABS computer booking systems	
	2 telephones	
	1 microphone unit all call/select zone	
	1 radio base unit	
	1 pool alarm base unit	
	Entrance call button indicators	
	Alarm indicator disabled toilets	
	Assistance call indicator of control barrier	
	CCTV monitor and keyboard	
	Personal attack alarms for each workstation	
	Lighting switching for entrance lobby and reception area	
Base Units:	PA rack unit	
	Cash transfer entry point	
	Lockable safe drawer for each workstation	
	2 lockable drawers for sale item stock	
	1 lockable deep drawer for lost property	
	Lockable rack for equipment	
Design	Uncluttered appearance	
	Centre section of reception to be 'dropped', serving wheelchair users and children	
	High quality materials	
	Security screen to be agreed	
	Secure access to reception	
Remarks		

Table 8.6 Typical room data sheet of a multi-purpose function room

Room data sheet

Activity Space	Multi-Purpose Function Room		No. 16.1
Occupants	Crèche: 16 children + 2/3 adults		
	Meetings/Functions: 4-30 persons		
Floor Area	m²	Min 43.5	
Ceiling Height	m	2.6-2.8	
Floor Finish	Heavy duty contact carpet; skirting boards to wall		
Wall Finish	Painted lined blockwork, glazed screen to circulation area		
Ceiling Finish	Suspended ceiling system		
Air Conditioning	Temperature	C°	min 17
	Air Changes/Hour		H&V to cope with high and low occupancy levels
Visual Conditions	Illuminance	Lux	300, CIBSE guide
	Daylight	Desirable	
	Lighting	Recessed fittings, local control	
Sound conditions	Room must be suitable for quiet activities		
Other Requirements	Internal telephone outlet, wall mounted		
	2 twin 13 amp sockets, child proof		
	PA System	Voice only	
Fittings	Blind to screen for black out		
Note	No sharp edges anywhere in room		
Safety Conditions	As required		
Access	From circulation area		
Remarks			

FPS is a new technique in VM and is an additional step to usual VM exercises. It is applied after the functions have been identified, defined and weighted. It removes perceived restrictions and gives a precise definition of the needs, which the product or service must need to be relevant.

Examples of a project FAST diagram and FPS are illustrated in Figure 8.7, Figure 8.8, Table 8.7, and Table 8.8.

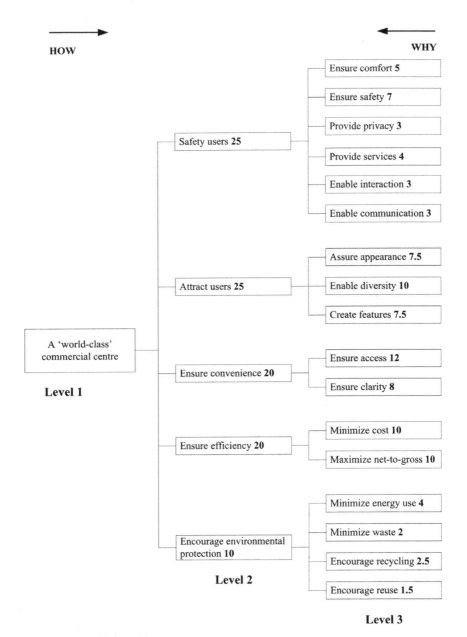

Figure 8.7 Example of the Project FAST diagram for a commercial centre with weighting assigned to functions

Adapted from Shen *et al.*, 2004.

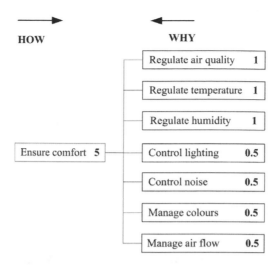

Figure 8.8 Example of the FPS application for 'regulating air quality'
Adapted from Shen *et al.*, 2004.

Table 8.7 Description of the acquired performance

Criteria	Level	Flexibility
Well-located air inlets	Avoid close proximity of outdoor intake to sources such as garages, loading docks, building exhausts, outside construction projects.	F1
Efficient circulation	Minimum ventilation (i.e. the introduction of fresh air to replace stale air): (1) about 0.5 to 3 air changes/hour depending upon density of occupants; (2) values per occupant range from 5 to 25 litres/sec/person (Baker & Steemers, 2000).	F2
	Air movement to cool heat sources: (1) average air velocity during winter not to exceed 30 feet/minute[1] (fpm); (2) average air velocity during summer not to exceed 50 fpm.	F2
Minimal airborne contaminants	High efficiency filter to be used for HVAC* system to remove bacteria, pollen, insects, soot, dust, and dirt (ASHRAE** dust spot rating of 85% to 95%) (EPA, 2001).	F1
	Areas from which fumes need to be extracted must be maintained at a lower overall pressure than surrounding areas, and be isolated from the return air system so that contaminants are not transported to other parts of the building.	F0
Allow for individual control	Local control system to modulate airflow.	F0
	Control switches to be conveniently located and properly instructed.	F0

Adapted from Shen *et al.*, 2004.
* HVAC – Heating, ventilating and air conditioning
** ASHRAE – The American Society of Heating, Refrigerating and Air-Conditioning Engineers

Table 8.8 Scale of flexibility in the FPS

Level	Description
F0:	The criterion is an absolute must, not negotiable, all effort must be made to meet this level, whatever the cost.
F1:	The criterion is a must if at all possible, no discussion unless there is a very good reason.
F2:	The criterion is negotiable, hope this level is reached, ready to discuss.
F3:	The criterion is very flexible, this level is proposed but is open to any suggestion.

Adapted from Shen *et al.*, 2004.

Action Plan

Workshops are the seats of action, and some form of action will arise from the discussions held therein. The 'action plan, itself, is a summary document that is usually incorporated or appended to the executive summary of the workshop report and describes the action in detail, the members of the team best suited to take the actions (whether present at the workshop or not), and the date by which the actions are to be completed' (Kelly *et al.*, 2004). Members of the workshop team will be the ones to proceed with all of the items identified in the action plan if the ACID Test has been correctly carried out.

The action plan is thus included in the project execution plan and the team's actions are reviewed at future design team meetings.

Table 8.9 Example of an Action Plan

Ref.	Action Description	By	When
1.	Prepare the room data sheets.	Ann Yu	1/3/2012
2.	Prepare the functional performance specification for building services installation.	Q.P. Shen	1/3/2012
3.	Prepare the project brief for circulation and agreement.	John Kelly	7/3/2012

8.10 References

BSI (2000). *Value management*, BS EN 12973:2000.

Duerk, D.P. (1993). *Architectural programming: information management for design*. USA: Van Nostrand Reinhold Company.

Construction Industry Board (1997). *Briefing the team: a guide to better briefing for clients*. London: Thomas Telford Ltd.

Kelly, J., Male, S., and Graham, D. (2004). *Value management of construction projects*. Oxford: Blackwell Science Ltd.

Masson, J. (2001). *The use of functional performance specification to define information systems requirements, create RFQ to suppliers and evaluate the supplier's response.* Conference paper of the SAVE Annual Congress, May.

Shen, Q.P., Li H., Chung, J., and Hui, P.Y. (2004). A framework for identification and representation of client requirements in the briefing process, *Construction and Management Economics,* 22(2), 213–221, Feb.

9 Performance Measurement of VM

Zuhaili Mohamad Ramly and Thomas G.B. Lin

9.1 Introduction

This chapter introduces the performance aspect of VM studies, which is important for determining how well the process worked and whether or not outputs from a study were satisfactory. Performance measurement is a learning process that provides feedback for future VM studies.

9.2 Learning Objectives

Upon completion of this chapter, you should be able to:

1. Explain the principles of performance measurement
2. Describe how to assess VM performance
3. Identify the indicators used to measure the performance of a VM study

9.3 Concept of Performance Measurement

Definition of performance measurement

The language in the field of performance measurement (PM) complicates the subject because it is so confused (Neely *et al.*, 2005). During the development of PM many terms appeared which were frequently used, such as, performance management, performance evaluation, performance measures, performance metrics, key performance indicators (KPIs), and critical successful factors (CSFs).

Neely *et al.* (1995) defined PM as the process of quantifying effectiveness and efficiency in actions by using a set of metrics. Bitichi *et al.* (1997) defined PM as the process of determining how successful organisations have been in attaining their objectives, while Waggoner *et al.* (1999) defined it as the process of monitoring the performance, identifying the areas that need attention, enhancing motivation, improving communications, and strengthening accountability. In general, PM involves evaluation of performance against a defined goal by identifying deviations from the expected results of pre-determined indicators (Latiffi *et al.*, 2010).

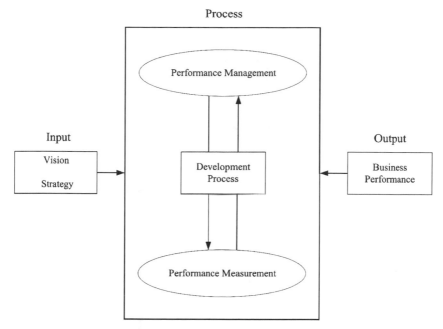

Figure 9.1 The performance management and performance measurement process
Adapted from Kagioglou *et al.*, 2001.

Functions and reasons for performance measurement

According to Amaratunga and Baldry (2002), lack of appropriate PM can act as a barrier to change and improvement. They claimed that measurement provides the basis for an organization to:

- Assess how well operational activities progressing
- Identify areas of strength and weaknesses
- Decide on future initiatives
- Attract future investment, increase share value and attract high calibre employees
- Learn from their past experience and perform better in the future, thus creating a good platform for learning culture of the organisation
- Understand and effectively satisfying the customer's expectations and needs
- Estimate and justify resources allocation
- Motivate the employees to improve performance

9.4   Performance Measurement Model

General concept of performance measurement

It has been a long time since PM was first adopted to assess the success of an organization. Before the 1980s, attention was mainly paid to financial accounting

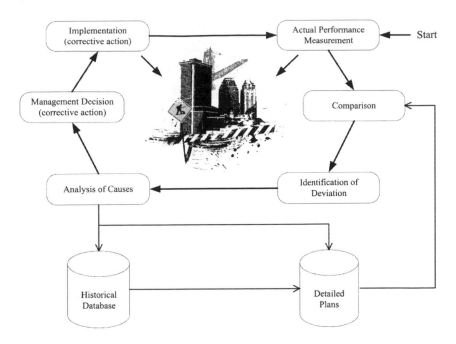

Figure 9.2 The performance control cycle

Adapted from Navon, 2007.

systems. In the 1980s, some researchers (Kaplan, 1983; Johnson and Kaplan, 1987) found that it is insufficient to measure an organization's performance merely using traditional financial accounting systems, whereby indicators, such as return on investment, return on assets, productivity, cash flow and so on, were calculated exactly to measure the financial performance of an organization.

As the world has developed from the industrial age to the information age, companies have to deal with global competition and rapidly changing techniques. However, PM based on financial accounting failed to provide the necessary information on how to improve competence in such a severe competitive environment. Some limitations of traditional measures are that they:

- Are historical in nature, reporting the outputs of previous months, quarters or years, which may be too old to base any decisions on (Dixon *et al.*, 1990).
- Give little indication of the relationship between work done in the present and performance in the future.
- Encourage focusing on short-term profits but not long-term strategies.
- Try to quantify performance and other improvement efforts solely in financial terms.
- Hinder innovation by placing too much emphasis on cost reduction and productivity, which inhibits the operational members from taking the initiative to produce something better.

- Are internally rather than externally focused, with little regard for competitors and customers.
- Lack a management information system (MIS) to ease the process of data storage, sorting and retrieval.

Evolution of performance measurement

Shifts to the traditional business paradigms caused by the introduction of new methods and techniques aimed at improving performance have led to rethinking the effectiveness of PM systems (Kagioglou *et al.* 2001). To increase the viability of PM, indicators outside of the financial realm have been added. The late 1980s and early 1990s saw the development of several integrated PM frameworks. Three of these became very influential: the strategic measurement analysis and reporting technique (SMART) (Cross and Lynch, 1988); the performance measurement questionnaire (PMQ) (Dixon *et al.*, 1990); and the balanced scorecard (BSC) (Kaplan and Norton, 1996). These frameworks consider both financial and non-financial measures at all levels of the organization. While they are not perfect, and not a complete solution to improving PM, they do add a new perspective to PM by going beyond traditional accounting systems.

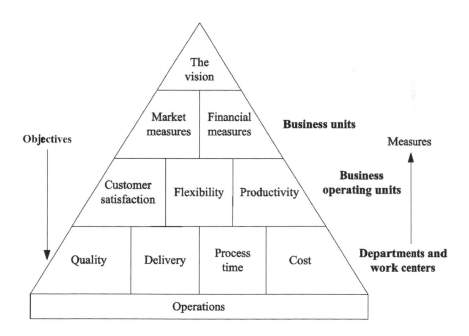

Figure 9.3 A framework for performance measurement system design

Adapted from Ghalayani and Noble, 1996.

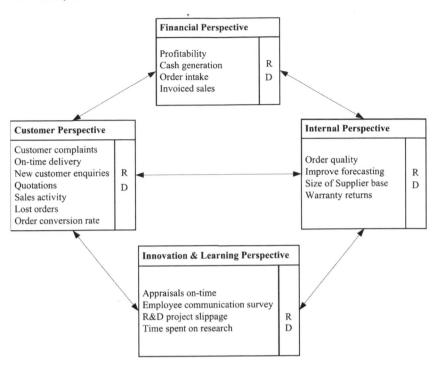

Figure 9.4 An example of balance scorecard of an organisation

Adapted from Bourne *et al.*, 2000.

Issues pertinent to performance measurement

Perrin (1998) found evidence that the PM approach is repeating the mistakes of the past with regard to the use of indicators for determining performance. Perrin criticised the use of indicators in PM from the following perspectives:

- **Definition of term used in measurement** – Indicators used in measurement processes can be interpreted in various ways across the different levels of an organisation and can create misunderstandings by different individuals within the organisation. Hence, it is crucial to provide a clear and unambiguous definition of common terms such as input, process, output, outcome, indicator, measure, efficiency, effectiveness, and productivity. Training, assistance and other form of support by management must be provided to the individuals directly involved with measuring and recording the data (Perrin, 2011).

- **Goal displacement** – There can be a tendency for indicators to become the objective when too much emphasis is put on the process in achieving the targeted indicators, without actually improving the overall outputs. Moreover, heavy dependency on quantitative indicators can disguise and mislead rather than inform what is really happening. There is also potential for data to be manipulated to make it look good, sound reasonable and meet the goals (Kueng and Krahn, 1999). As a result, performance indicators are frequently distorted in the direction of programs, diverting attention away from what the outputs should be.

- **Meaningless and irrelevant indicators** – Meaningless and irrelevant indicators are often developed and used (Perrin, 1998; Bernstein, 1999) because they were chosen in haste without sufficient consideration (Davies, 1999). There is also the possibility of rejection by top management of indicators that they did not intended to have according to their interpretation. Hence, the development of indicators requires commitment from top management throughout the design, implementation, and eventual use of the data (Nudurupati *et al.*, 2011). Kueng and Krahn (1999) suggested that working on process goals is a good starting point to determine the right indicators. In line with this, Perrin suggested the bottom-up approach for the selection of indicators, to allow individual staff to understand the purpose of the collected data.

- **Cost saving vs. cost shifting** – According to Perrin (1998), indicators used in PM are invariably short term in nature. Hence, short-term benefits and outputs may result in requirements and increased costs over the longer term that may shift into other future costs.

Even though Perrin (1998) was very critical, he nevertheless offered strategies to minimise the risk of misusing performance indicators by considering who is likely to make use of the indicators, identifying potential misinterpretations, developing indicators that are general enough to minimize actual misuse, and testing indicators in advance to anticipate potential misapplications. Since organisations will continue for a long period of time, performance indicators must be frequently reviewed, revised and updated at regular intervals to match current needs, circumstances, opportunities and priorities, and to keep them in line with the organisation's goals and objectives (Bernstein, 1998; Nudurupati *et al.*, 2011).

Performance measurement model in construction

Traditional PM in construction was oriented based on the end product as a facility and the process involved in constructing it. The 'iron triangle' used to measure the performance known as cost, time and quality. However these measures can only be determined after the project has been completed. This is what Kagioglou *et al.* suggested as 'lagging' indicators as they do not provide the true performance of the various factors that may affect the project. Thus the performance of a project should be measured in a more comprehensive way. Kumaraswamy and Thorpe (1996) suggested adding client satisfaction, project team satisfaction, technology (trans-fer), environment (friendliness), and health and safety to the success criteria of projects. The KPI framework (KPI working group, 2000) created seven key per-formance indicators: time, cost, quality, client satisfaction, change orders, business performance, and health and safety.

Because of the project based nature of construction business units (Liu and Walker, 1998), it is even more important to perform project performance because each project has unique features (Bassioni *et al.*, 2004). This uniqueness makes it very difficult to develop a generic framework to measure the performance of a project. If such a generic framework were used, it would struggle under the different assessments and perspectives of stakeholders, such as the client, end user, consultants, and contractors.

9.5 Performance Measurement of VM studies

While VM continues to be used in many projects to help achieve better value for money, specifically in the challenging environment of budget constraints, safety issues and environmental concerns, VM studies still face pressure due to limited time and resource allocation (Shen and Liu, 2003). A good performance measurement system of VM studies is therefore important to improve the time and resource allocation afforded to VM studies.

Critiques of VM studies

Although VM has been applied in construction for about 40 years and has obtained a high reputation, criticisms of it have never ceased. A number of authors criticize and question the effect of VM exercises. Typical critiques of VM in the construction industry include:

- It is normally difficult to assemble key stakeholders for 40 hours of workshop and retain their attention throughout this period. Moreover, considerable time is needed by the design team for reviewing VM proposals and re-designing after the workshop. This is an even more challenging situation where the project schedule is very tight.
- Researchers have advocated that VM should be implemented as early as possible to maximize its influence on the project formulation (Green, 1994; Dell'Isola, 1997). However, the most common point for VM intervention in practice is 35 percent of the way through the design development process, since any changes to the original design are more easily introduced, costing data is more readily available in the form of cost estimates, and savings can easily be identified.
- VM intervention at 35 percent of the way through the design development is essentially a design audit (Palmer *et al.*, 1996). The adversarial attitude of the original design team is not easily eliminated. Many designers have argued that in a short period of time the VM team could not be expected to fully understand the project in comparison to the existing design team.
- The design liability of VM proposals is a thorny issue in VM applications. Whether the VM team or original team takes the design liability for any recommendations implemented is debatable, although the design team determines whether to accept or reject any proposals recommended by the VM team.

- A number of influential VM authors have given a strong emphasis to function analysis. However, function analysis sometimes is not clearly defined during the workshop process even though it the part of the VM methodology that distinguishes VM from cost cutting exercises.

Traditional performance measurement in VM studies

In the past, PM of VM studies focused on improved functions and cost savings achieved by implementing the proposals of VM studies. Kelly and Male (1993) and Palmer *et al.* (1996) introduced a PM framework that only focused on cost reduction (Shen and Liu, 2003; Lin *et al.*, 2004). It was claimed that many other aspects, such as the clarification of clients objectives, improvement of communication among stakeholders and the acceleration of decision-making should be considered (Norton and McElligott, 1995; Palmer *et al.*, 1996; Shen and Liu, 2003). Apparently, most of the process-related performance indicators are seldom considered, which hinders VM practitioners from knowing the relationship between the processes and outputs.

Development of performance measurement in VM studies

In a benchmarking study performed by Male *et al.* (1998a), ten critical success factors (CSFs) for VM studies were identified. The research showed that client satisfaction was a good indicator of the performance of VM in a project, and that relevant project data collected after VM was applied could be added to give an overall indication of the VM performance. A literature review of factors affecting the success of VM studies by Shen and Liu (2003), identified 15 CSFs. These CSFs can be grouped into four clusters: 'value management team requirements', 'client influence', 'facilitator competence', and 'relevant department impact'. Since each VM study is unique, based on the unique parent project, very flexible measurement systems are needed. At the same time they must be fast to implement and cost-efficient because of the time and resource limitations of VM studies.

Evolution of performance measurement in VM studies

Generally, VM studies are process-based and conducted through a structured methodology to achieve specific goals and objectives. The development of PM frameworks provides guidance as to what to measure, how to measure, and also how to make the best use of measures (Nudurupati *et al.*, 2011). However, it is found that process-related performance is seldom considered when measuring performance. Hence calls for measurement systems with detailed but flexible indicators to represents what is really happening during VM studies, along with expected outputs.

Lin (2009) proposed a framework for measuring the performance of VM studies in construction by taking an approach in line with the nature of VM studies. To overcome the deficiencies of other PM systems, the three major concerns in developing the framework were:

- The framework should measure both the process of the VM study and the final output.
- The data collection and processing methods of the framework must be fast to provide feedback for corrective measures in a timely fashion.
- The non-financial indicators that are crucial to the success of the VM study should be included in the PM framework.

Based on the nature of VM studies, the process of construction projects and the principles of measuring performance, Lin *et al.* (2005) conducted a literature review and examined VM reports to identify thirteen major factors that have an impact on the performance of a VM study (see Figure 9.7). With the proposed PM framework in mind, the indicators used to quantify the efficiency and effectiveness of VM processes and outputs were determined. Each CSF should have a few KPIs that can be measured and quantified for easier comprehension.

From a long list of potential performance indicators, eighteen were chosen as the key performance indicators (KPIs) and grouped into three categories: predicting indicators, process performance indicators, and outcome performance indicators.

9.6   Performance Indicators of VM studies

KPIs can be defined as measurers that are used to determine the performance level of a particular process or system. From Figure 9.8 it can be seen that KPIs include process indicators, tangible outcome indicators and intangible outcome indicators. The measurement framework including these KPIs takes into consideration intangible performance, which is not taken into consideration by previous frameworks.

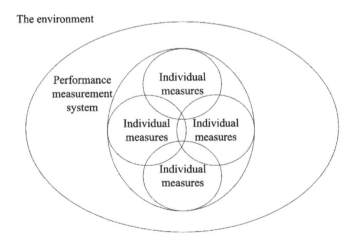

Figure 9.5 A framework for performance measurement system design

Adapted from Neely *et al.*, 1995.

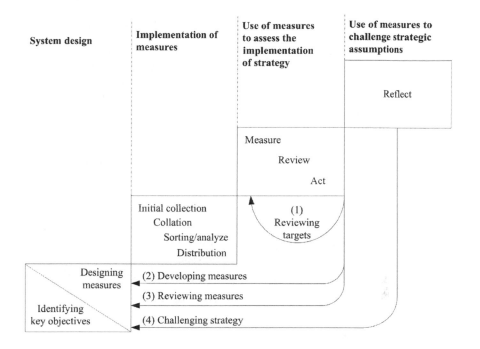

Figure 9.6 Phases in developing the performance measurement system/framework

Adapted from Bourne *et al.*, 2000.

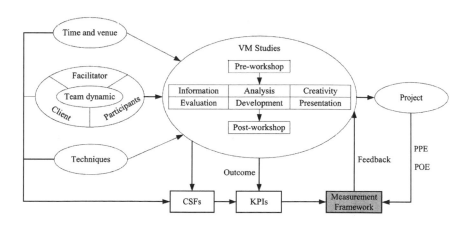

Figure 9.7 Theoretical framework for performance measurement of VM studies in construction

Adapted from Lin *et al.*, 2005.

Predicting indicators

The predicting indicators are among the critical factors that may influence the performance of VM studies. Among others, Lin (2009) found that clear objectives of the workshop, client participation, client's support, disciplines of participants, qualification of facilitators, and relevant department's support are critical indicators prior to the commencement of the workshop. While Interaction between participants in each phase, client's objectives clarified, project assumptions clarified, and primary function identified are critical throughout the workshop process implementation during the workshop stage. There are several finding on similar themes by Palmer *et al.* (1996), Male *et al.* (1998), Simister and Green (2002), and Fong *et al.* (2001).

Process performance indicators

Process performance indicators as found by Lin (2009) includes the background information collected during the pre-workshop stage, client's objectives clarified, interaction between participants in each phase, primary function identified, and project givens/assumptions clarified.

Outcome performance indicators – Tangible

Tangible indicators are something that is measurable. As an output of the workshop, *Quality of the report* and *percentage of action plan carried out* are easily determined to indicate the performance. Another indicator would be *the quality of the project* once it is finished. A well-written report will help the client grasp the processes and output of the VM study. It also guides the client's decision making in relation to any recommendations from the VM study. The benefits of the VM study to the entire project can be realized only if the follow-up actions recommended by the study are carried out. The KPI 'Improving the project quality' differentiates VM from mere cost reduction techniques.

Outcome performance indicators – Intangible

Intangible indicators are inherently hard to measure, and include such things as: identifying and clarifying client requirements, accelerating the decision-making process, improving internal communication, increasing understanding among stakeholders, and improving client satisfaction. These KPIs are all missing indicators that indicate the outputs of the workshop. Identification of these indicators is consistent with earlier research findings from Male *et al.* (1998) and Lin and Shen (2007).

9.7   Conclusion

VM offer benefits to the construction projects and to the relevant stakeholders. As a management tools, VM involved processes at different stage; pre-workshop,

workshop and post-workshop. Those processes require specific tools to achieve the workshop objectives and active participation by the stakeholders under the facilitation by the VM facilitator.

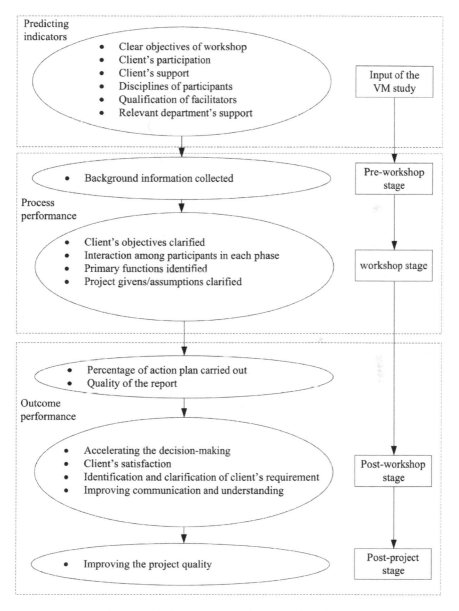

Figure 9.8 Key Performance indicators (KPIs) for measuring the performance of VM studies in construction

Adapted from Lin, 2009.

In light of the limited time, resources and money invested in order to conduct a VM workshop, measuring the performance of VM studies is becoming more important. It provides an indication on how well the workshop was conducted and how VM influences and benefits a project. Flexible PM framework is necessary in order to allow the smooth process of measuring the performance. The identification of appropriate indicators is crucial to ensure that it represent the overall performance of the VM workshop.

9.8 References

Amaratunga, D. and Baldry, D. (2002). Moving from performance measurement to performance management, *Facilities*, 20(5/6), 217–223.

Bassioni, H.A., Price, A.D.F., and Hassan, T.M. (2004). Performance measurement in construction, *Journal of Management Engineering*, 20(2), 42–50.

Bernstein, D.J. (1999). Comments on Perrin's 'Effective use and misuse of performance measurement', *American Journal of Evaluation*, 20(1), 85–93.

Bitichi, U.S., Carrie, A.S., and McDevitt, L.G. (1997). Integrated performance measurement system a development guide, *International Journal of Operations & Production Management*, 17(5), 522–534.

Bourne, M., Mills. J., Wilcox, M., Neely, A., and Platts, K. (2000). Designing, implementing and updating performance measurement systems, *International Journal of Operations & Production Management*, 20(7), 754–771.

Cross, K.F. and Lynch, R.L. (1988). The 'SMART' way to define and sustain success, *National Productivity Review*, 8(1), 23–33.

Davies, I.C. (1999). Evaluation and performance management in government, *Evaluation*, 5(2), 150–159.

Dell'Isola, A.J. (1997). *Value Engineering: Practical Application: for Design, Construction, Maintenance, and Operations*. Kingston Mass: RSMeans.

Dixon, J.R., Nanni, A.J., and Vollmann, T.E. (1990). *The New Performance Challenge – Measuring Operations for World-Class Competition*. Homewood, IL: Dow Jones-Irwin.

Fong, P.S., Shen, Q.P., and Cheng, W.L. (2001). A framework for benchmarking the value management process, *Benchmarking: An International Journal*, 8(4), 306–316.

Ghalayini, A.M. and Noble, J.S. (1996). The changing basis of performance measurement, *International Journal of Operations & Production Management*, 16(8), 63–80.

Green, S.D. (1994). Beyond value engineering: SMART value management for building projects, *International Journal of Project Management*, 12(1), 49–56.

Kagioglou, M, Cooper, R., and Aouad, G. (2001). Performance management in construction: a conceptual framework, *Construction Management and Economics*, 19(1), 85–95.

Kaplan, R.S. and Norton, D.P. (1996). *The balanced scorecard: translating strategy into action*. Boston: Harvard Business School Press.

Kelly, J. and Male, S. (1993). *Value management in design and construction: the economic management of projects*. London: E & F.N. Spon Press.

KPI Working Group (2000). *KPI Report for the Minister for Construction, Department of the Environment*, Transport and the Regions, London.

Kueng, P. and Krahn, A.J. (1999). Building a process performance measurement system: some early experiences, *Journal of Scientific and Industrial Research*, 58(3/4), 149–159.

Kumaraswamy, M.M. and Thorpe, A. (1996). Systematizing construction project evaluation, *Journal of Management in Engineering*, 12(1), 34–39.

Latiffi, A.A., Carrillo, P.M., Ruikar, K.D., and Anumba, C.J. (2010). *Current performance measurement practices – studies in the United Kingdom (UK) and Malaysia*, Proceeding of 18th CIB World Building Congress, Salford, UK, 303–315.

Lin, G.B. and Shen, Q.P. (2007). Measuring the performance of value management studies in construction, *Journal of Management in Engineering*, 23(1), 2–9.

Lin, G.B., Shen, Q.P., and Fan, S.C. (2004). *A framework for performance measurement of value management studies in construction*, Proceedings of the 3rd International Conference on Construction and Real Estate Management, Hong Kong: China Architecture & Building Press, 307–311.

Lin, G.B. (2009). *Measuring the performance of VM studies in construction*, Unpublished PhD thesis, The Hong Kong Polytechnic University.

Liu, A.M.M. and Walker, A. (1998). Evaluation of project outcomes, *Construction Management and Economics*, 16(2), 209–219.

Male, S., Kelly, J., Fernie, S., Gronqvist, M., and Bowles, G. (1998). *The value management benchmark: research results of an international benchmarking study*. London: Thomas Telford.

Navon, R. (2007). Research in automated measurement of project performance indicators, *Automation in Construction*, 16(2), 176–188.

Neely, A., Gregory, M., and Platts, K. (1995). Performance measurement system design: a literature review and research agenda, *International Journal of Operations & Production Management*, 15(4), 80–116.

Norton, B.R. and McElligott, W.C. (1995). *Value management in construction: a practical guide*. Hampshire, UK: McMillan Press.

Nudurupati, S.S., Bititci, U.S., Kumar, V., and Chan, F.T.S. (2011). State of the art literature review on performance measurement, *Computers & Industrial Engineering*, 60(2), 279–290.

Palmer, A., Kelly, J., and Male, S. (1996). Holistic appraisal of value engineering in construction in United States, *Journal of Construction Engineering and Management*, 122(4), 324–328.

Perrin, B. (1998). Effective use and misuse of performance measurement, *American Journal of Evaluation*, 19(3), 367–379.

Perrin, B. (1999). Performance measurement: does the reality match the rhetoric?, *The American Journal of Evaluation*, 20(1), 101–111.

Perrin, B. (2011). What is a result/performance-based delivery system? An invited presentation to the European parliament, *Evaluation*, 17(4), 417–424.

Shen, Q.P. and Liu, G.W. (2003). Critical success factors for value management studies in construction, *Journal of Construction Engineering and Management*, 129(5), 485–491.

Simister, S.J. and Green, S.D. (1997). Recurring themes in value management practice, *Engineering, Construction and Architectural Management*, 4(2), 113–125.

Waggoner, D.B., Neely, A.D., and Kennerley, M.P. (1999). The forces that shape organisational performance measurement systems: an interdisciplinary review, *International Journal of Production Economics*, 60–61, 53–60.

10 VM Case Studies

Geoffrey Q.P. Shen

10.1 Introduction

This chapter introduces two real life construction projects as case studies for the implementation of value management (VM).

10.2 Learning Objectives

Upon completion of this chapter, you should be able to:

1. Demonstrate how VM is applied to real life construction projects
2. Identify the key issues to be addressed when implementing VM in such projects

10.3 Case 1: Rehabilitation of Housing Estate

Executive Summary

This VM study was conducted as a means of assisting the project team and other stakeholders to take a step along the value improvement path for works relating to the rehabilitation of an old housing estate. The VM workshop addressed all major aspects of the design and construction of the rehabilitation work with its primary objective being to:

> Create a structured forum whereby views from all stakeholders on what and how rehabilitation work should be implemented can be discussed openly in order to reach a consensus on the best value solutions for the rehabilitation project.

The workshop objective was confirmed as appropriate by the team as a whole at the commencement of the workshop. The project was at the concept design stage when the VM study was conducted and all participants were focusing on adding value by addressing the above objective in a structured workshop. It should be noted that the overall objective of the VM workshop was not necessarily to make final decisions but to establish the parameters by which best choices could be made. The primary objective was achieved by:

- Developing/confirming the project objectives, stakeholders' value system, and performance criteria.
- Understanding the problem situation, the areas of risks, concerns, and opportunities to the project.
- Developing a shared understanding of the opportunities we have in achieving the project objectives.
- Focusing upon the necessary functions that we identify as the key value drivers.
- Identifying potentially viable options to address the critical issues identified during the workshop.
- Agreeing upon the criteria to be utilised for evaluation of the identified options.
- Carrying out an initial evaluation of the identified options based upon the available information.
- Developing a detailed action plan to further develop/implement our responses to the key issues identified at this workshop.

The VM study was conducted in accordance with the Australian and New Zealand Standard AS/NZS 4183:1994, which includes five phases: Information, Analysis, Creativity, Evaluation, and Development. Tasks in each phase of the workshop were as follows.

Information Phase:

- Introduce project parameters, objectives, and scope
- Discuss and agree workshop objective and scope
- Define and redefine givens and assumptions

Functional Analysis Phase:

- Analyse the functions of the chosen facilities
- Construct functional hierarchy/FAST diagrams

Creativity Phase:

- Generate alternative solutions for the identified functions
- Identification of value-improvement alternative solutions

Evaluation Phase:

- Grouping ideas to re-define/create new options
- Discuss and agree evaluation criteria
- Evaluate options created in the previous phase

Development/Action Phase:

• Agree follow-up actions

The one-and-a-half-day workshop was conducted on two separate days, with one week in-between (the detailed rundown of the workshop is shown in Appendix A). This was to give the team members sufficient time to obtain additional information, to investigate possible solutions, and to follow up proposed actions before meeting again on the second day of the workshop. This arrangement proved to be very successful. Stakeholders' views were solicited and a consensus was reached, which established the direction for the design and construction work of the project.

The Action Plan represents the team's deliberation on actions to be followed up, which will further advance the outcomes of the workshop, assist the formation of detailed design and construction proposals, and help with implementation of the project.

In summary, the workshop provided an open forum to address key issues and develop ideas and opportunities to address these issues. The workshop was also utilised to improve the working relationships that evolved through the desired team approach to this project. The workshop objective was achieved and expressed in the form of actions and options reflecting improvements in value and the key actions required; these are presented in the workshop outcomes section of this chapter. During the workshop, all participants contributed to and were exposed to a high level of information, establishing the requirements of the key stakeholders with reference to the functional requirements of the project.

Workshop Outcomes

The VM study consisted of a structured process of achieving the specific workshop objective. As a result of the approach, workshop participants shared a large amount of information and generated a large number of alternative solutions. The inform-ation was utilised in determining the focus and specific direction of the workshop, and was the basis for developing the required outcomes.

The workshop was conducted in a function room of a clubhouse (Figure 10.1) in order to provide an environment that would facilitate the flow and exchange of large amounts of information. The following representatives of the stakeholders of the proposed rehabilitation project participated in the workshop:

Director (Property Development)	Architectural Assistant
General Manager (Planning & Development)	Director/RSE
Senior Manager (Planning & Development)	Assistant Engineer
Contract Manager	Senior Engineer
Project Manager	Project Director
Quantity Surveying Manager	Quantity Surveyor

Quality Assurance Manager	Senior Project Building Engineer
Building Services Engineer (Quality Assurance)	Project Manager – Retrofitting
Architect (Quality Assurance)	Senior Project Manager
General Manager (Maintenance)	Senior Planning Engineer
General Manager (Property Management)	Project Quantity Surveyor
Manager (Property Management)	Site Agent
Senior Manager (Building Services)	Contract Manager (Building Renovation)
Manager (Maintenance – Building)	Assistant Project Manager (Engineering)
Manager (Maintenance – Building Services)	Building Services Manager
Quantity Surveyor (Contract Management)	General Foreman
Project Superintendent (Clerk-Of-Work)	Workshop Facilitator
Project Running Associate	Workshop Secretary

Information Phase

A brief introduction to VM was made to the team of participants at the commencement of the workshop. Several introductory presentations were also given in order

Figure 10.1 Setting of the VM workshop

to provide background information to representatives of the key stakeholders of the project. The presentations covered the following issues:

- Project objectives
- Design proposal
- Structural design
- Preliminary construction proposal

The workshop participants identified a number of issues, concerns and opportunities that could favourably or unfavourably affect the project. Examples of these critical issues are: money and cost, political sensitivity, and buildability.

Project Givens

The workshop participants identified a list of givens for the project, which are things that we *know* to be true, such as:

- Project will go ahead
- Site boundary
- Occupied estate

Project Assumptions

The workshop participants generated a list of assumptions for the project, which are things that we *think* to be true, such as:

- Access to tenants' flats
- Working area outside site boundary
- Site access

Selected Key Issues for Team Focus

A large number of issues were identified by the stakeholders during the workshop, which are relevant to this rehabilitation project. The participants ranked the issues in order to identify the key issues on which the team should focus.

Cluster	Key issues to be addressed
1	Suspension of services e.g. water, electricity, gas Disturbance to normal life
2	Safety to both tenants and workers Transportation of materials Access for workers Traffic control and arrangement for tenants

3	Buildability
	Decanting facilities/arrangement
4	Phasing – may affect cost & tenants' support
5	Money and cost
	Overall duration of the project

Function Analysis Phase

It is important to establish a clear understanding of the purpose of the project and how the success of the project will be measured. Three primary project objectives were identified, and all of them serving the mission of 'fulfilling social responsibility'. Project functions were also identified and presented in the form of a functional hierarchy to clarify the understanding of the project objectives among participants. Figure 10.2 shows the functional hierarchy.

Creativity Phase

A large number of value improvement proposals were generated to address areas identified as having the maximum potential for value improvement. The team reviewed *Can We* options and those assessed as realistically achievable were then grouped into themes and value improvement proposals known as *We Can* statements. Options that were classified as remotely possible were not addressed further in the workshop.

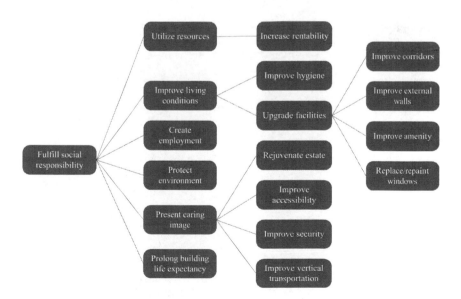

Figure 10.2 Functional Hierarchy of the Rehabilitation Project

Evaluation Phase

Six key criteria, identified by the team of stakeholders as reflective of their key values, were used to assess all proposals developed during the workshop. The weightings of these criteria were obtained through a pair-wise comparison of values by participants. They are as follows:

	Criteria	Weighting
A	Cost (initial and operational)	4
B	Corporate image	7
C	Customer satisfaction	6
D	Time/duration	1
E	Quality of product	3
F	Safety (to both tenants and site workers)	10

Following a systemic process of information sharing, idea generation, evaluation, development, and consensus building at the workshop, a number of proposals were selected for further development. Actions proposed during the first day of the workshop were dealt with before the second day of the workshop.

Development Phase

Transfer Lift

Improving accessibility of the estate by providing a transfer lift

- Identify existing services/utilities in the proposed location
- Arrange surface diversion/by-pass provision
- Minimise suspension of services
- ......

Actions Arising from Day-One of the Workshop (examples)

Actions Required	Actions Taken and Team Consensus
Identify existing services/utilities in the proposed location	11kV cables and low V reserved cables, gas pipes and Cable TV facilities exist. Alternative design is feasible to carry out, but special attention should be paid to the construction of foundation piles.
......	......
	Lifts at Linear Blocks **Improving accessibility by providing a lift at each linear block**

- Carry out services diversion during the construction of the foundation of lift towers
- Maintain accessibility of the adjacent staircase by tenants
- Consider open-able hoarding for the lift tower construction
- ……

Actions Arising from Day-One of the Workshop

Actions Required	Actions Taken and Team Consensus
Investigate the feasibility of building an elevated temporary walkway over the slope for pedestrians	Three additional options were proposed: Option 1 – block the whole walkway during construction; Option 2 – provide working areas at both side of block, the walkway will be blocked temporarily upon request. Option 3 – partly closed according to the work progress.
……	……

Covered Walkway + Footbridge

Improving convenience by providing a covered walkway and footbridge for the residents

- Arrange permanent diversion of existing underground/exposed drainage
- Minimize column footings of covered walkway and services diversion by use of suspension construction system, need to explore alternative footbridge design
- Arrange temporary pedestrian diversion during construction/ Fence off working area to protect residents
- ……

Actions Required	Actions Taken and Team Consensus
Minimize column footings of covered walkway and services diversion by use of suspension system, need to explore alternative footbridge design	2 options: (1) whole base is footing, no facilities diversion is needed; and (2) traditional footing. Both designs can be used, depending on the underground conditions. The traditional one is easier to conduct service maintenance in the future.
……	……

Corridors and External Walls

Improve tenants' living conditions by making improvement to corridors and external walls

- Provide a rest room for tenants during construction
- Provide exhaust fan blowers during removal of internal finishes along corridors to reduce dust nuisance
- Carry out corridor improvement in two longitudinal phases, similar to the experience of similar projects
- ......

Actions Required	**Actions Taken and Team Consensus**
Have alternative amenity services during cut off period	The team feels that the best option is to provide a rest room during construction.
......	......

Improvement to Vacant Flats

Improve living conditions by making improvement to vacant flats

- Services diversion: temporarily cap off the branch pipes (drain and flushing) of vacant flats; only reconnect when the new work is complete.
- Minimize disturbance (approximately one month for removal of sunken slab and installing new pre-cast floor slab, excluding finishing works) to normal life.
- Fence off works area to protect tenants
- ......

Actions Required	**Actions Taken and Team Consensus**
Services diversion: temporarily cap off the branch pipes of vacant flats; only reconnect when the new work is complete.	Use the standard precast unit to minimize disturbance. It takes only a few hours to complete the task.
......	......

Action Plan

Following a systemic process of information sharing, idea generation, evaluation, development, and consensus building at the workshop, a number of actions were agreed upon by the team. This Action Plan represents a consolidation of the workshop outcomes and the return on investment in the VM Workshop will depend very much on the vigour with which the actions recommended in the plan are pursued.

The following section records actions proposed during the second day of the workshop. Actions proposed on day-one of the workshop were dealt with before the second day of the workshop and recorded as value improvement proposals during the creativity phase.

No	Action Required (examples)	By Whom	By When
	Workshop Report		
	Issue draft report to client representative for initial comments and feedback.	Facilitator	26/5
	Supply initial feedback on draft report to Facilitator.	XXXXX	1/6
	Issue revised draft report to client representative for circulation to workshop participants.	Facilitator	2/6
	Supply collective feedback on revised draft report to Facilitator.	XXXXX	10/6
	Revise and issue final report to client representative for circulation to all workshop participants.	Facilitator	13/6

......

10.4 Case 2: Collection and Storage of Surface Water

Executive Summary

This is the first VM workshop for the captioned project, focusing on the identification and evaluation of feasible and cost-effective options for maximising the collection of surface water. The workshop was organised on 15 June 2001 and was attended by 16 representatives of the project's major stakeholders, including the client department, the consulting firm, and relevant government departments.

The VM workshop followed a widely used structured procedure known as the VM Job Plan, which consists of information phase, analysis phase, creativity phase, evaluation phase, development phase, and presentation phase.

Following a brief introduction to the workshop programme and the VM methodology, the participants agreed a workshop contract and confirmed the following primary objective of the project's VM workshop:

> To explore and evaluate feasible and cost-effective options for maximising the harvesting of surface water

The information phase started with a presentation by the consultant on Information Paper No. 1 of the project. This was followed by comments and clarification from the participants. The scope of the workshop had been clearly defined through identification and clarifications of the givens and assumptions of the project.

It was agreed that this is one of several strategic studies initiated by the client department to investigate alternative ways of providing a supply of fresh water. The focus of this particular study was the collection and storage of surface water, and the comparison of this method with other alternatives, such as desalination of seawater, that would be conducted by the client department in due course.

During the analysis phase, the VM team identified and agreed that the primary functions of the project were: (a) harvest surface water and (b) maximise the harvesting, which serve three wider purposes: (1) support the current and future sustainable development, (2) provide alternative sources of water supply, and (3) reduce the risk of fresh water shortage.

During the creativity phase, the brainstorming technique was deployed to assist the multi-disciplinary team to generate alternative options of achieving the functions defined for the project. The team generated a total of 35 ideas for the functions. These ideas were classified into three categories: realistically possible (P1), remotely possible (P2), and fantasy (P3).

A total of 25 ideas were ranked as P1, they were grouped into five clusters and the team developed them further into integrated solutions for implementation. These five clusters of feasible ideas were:

A.   Improve the harvesting of surface runoff from the existing water gathering ground (WGG), water collection and storage facilities
B.   Extend WGG, water collection and storage facilities
C.   Harvest flood water
D.   Harvest storm water
E.   Utilise raw water from existing irrigation and recreation reservoirs etc.

During the evaluation phase, the team held a lengthy discussion on the criteria to be used in the evaluation of various options for the project, which represent major factors to be taken into consideration in the evaluation process. A total of 23 criteria were raised and agreed upon by the team, which were grouped in five categories: functionality, costs, environmental impact, regulatory and social issues, and additional benefits. The weightings for these criteria were also discussed by the team. The consensus was that the first four criteria should be treated with equal importance and that they are more important than the No. 5 criterion. It was agreed by the team that these selected criteria and their weightings would be used to evaluate the options/schemes to be generated by the consultants in the subsequent stages of the project.

During the development phase, an action plan was prepared and agreed upon by the team to follow up issues raised during the workshop. A total of 12 actions with specified time frames for completion would be undertaken by the consultants and other stakeholders concerned. A strong emphasis was given to further studies on the five clusters of realistically possible ideas.

At the end of the workshop, the VM team confirmed that the objective of this workshop had been achieved. New feasible and cost-effective options have been identified and evaluated. Consensus was reached among team members on a number of issues, such as criteria for the evaluation process and follow-up action to be taken after the workshop. The end product of the workshop is owned by the entire team. This ownership fosters the commitment of the stakeholders during the implementation of the solutions.

Having gone through the process, the team believed that it is of paramount importance for the consultants to build on the consensus developed among the stakeholders at the workshop and to follow up the recommendations made by the team to conduct a detailed analysis of the new options. It was suggested that the consultants should investigate further the possible implementation of the five clusters of ideas and their consolidated solutions, develop them into feasible schemes of water collection, storage and usage, and evaluate these schemes using the agreed criteria and weightings to arrive at the best value for money solutions.

By using the structured approach, it was anticipated that the final feasibility study would contain the best options for maximising the collection, storage, and usage of surface water.

Background Information

For many years local groundwater and surface resources have been insufficient to meet domestic, municipal and industrial demands. A reappraisal of the collection and storage of surface water is required to determine whether it can be made available as a reliable and cost effective water source.

Although a substantial proportion of lands already form part of direct and indirect catchment areas, the optimisation of collection cost versus yield at the time each scheme was developed may now favour extension of Water Gathering Grounds (WGG) into regions not previously harvested. Introduction of new materials and techniques may also favour extension of WGG. The areas for potential extension of WGG are shown in the Brief.

For the communities, the impacts of extending WGG or of modifying flood-pumping areas must be fully assessed. Although communities accept existing arrangements, imposition of further restrictions or indeed any change to the status quo requires careful consideration. Future arrangements must remain acceptable in terms of the environment, public health, visual impact, traffic and overall safety.

The objectives of the assignment are to: (a) Investigate the feasibility to maximising the collection and storage of surface water in City X; and (b) Assess the quantity and quality of surface water that can be collected and stored in the impounding reservoirs.

Workshop Objective and Methodology

It has been suggested that the feasibility study would benefit from a structured VM study, through which multiple and cross-disciplinary issues involving the key stakeholders could be reaffirmed and resolved. To conduct this task in a manageable and systematic manner, the consultants proposed that two VM workshops should be conducted. The first VM workshop was conducted focusing on collection of surface water, whereas the second workshop focused on maximisation of surface water storage and its integration with the collection system.

The objective of this workshop was to: 'Explore and evaluate feasible and cost-effective options for maximising the harvesting of surface water'. The team endorsed this objective at the beginning of the workshop.

The VM workshop was attended by major stakeholders in seeking out alternative solutions of achieving the necessary functions at the lowest possible overall costs, consistent with requirements for performance. In order to achieve the above workshop objective, the following tasks were undertaken during the workshop:

a)   Examine the feasibility and the practicality of options
b)   Develop evaluation criteria and weighting to compare the options
c)   Evaluate the options and reach a consensus on the preferred solution
d)   Consider implementation of the preferred option

The findings and recommendations of the workshop are presented in this VM report, which was incorporated into the consultant's report submitted to the Study Management Group of the client department.

The VM workshop followed a widely used structured procedure known as the VM Job Plan. As shown in Table 10.1, the plan consists of six phases: information, analysis, creativity, evaluation, development, and presentation.

Before the workshop, a pre-workshop meeting was arranged with the consultants of the project and the representatives from the Water Supplies Department, to discuss matters relating to the VM workshop.

To prepare members for this workshop, an 'Information Paper' was prepared and issued to all participants in advance of the date of the workshop to ensure adequate briefing, and to serve as a paper for discussion at the workshop.

Table 10.1 Structure and Procedure of the VM workshop

Phase	Purpose/Activity
Information	To provide the team with a thorough, common, and precise understanding of the project.
Analysis	To identify primary functions of the project and to clarify givens and assumptions of the project.
Creativity	To generate alternative solutions to achieve the necessary functions through a brainstorming exercise.
Evaluation	To set up criteria for evaluation and use them to evaluate and select the preferred alternative solutions.
Development	To discuss and elaborate the implementation of the preferred options and develop an action plan which lists necessary follow-up actions.
Presentation	Report of the VM workshop.

The VM workshop was attended by representatives of major stakeholders of the project, including the client department, the consultants, and a number of government departments concerned. Opportunities were taken to solicit and address critical issues and concerns, and to arrive at solutions acceptable to all stakeholders. It provided a comprehensive and holistic view of various aspects of the project and sought ideas and alternative solutions for possible value enhancement.

The facilitator's role at the workshop was primarily to facilitate and manage the process and prepare a report on behalf of the team. His main tasks were to provide important stimulus to encourage ideas from all the participants, maintain the momentum of the group dynamics, recognise and overcome potential blocks that prevent the flow of innovative ideas, and establish an open forum where all participants' contributions are valued equally.

Process and Participants of the Workshop

A detailed programme for this workshop is shown in Appendix B. In essence, the VM workshop consists of the following phases and activities:

Information Phase:

- Project parameters, objectives, and scope
- Workshop objectives and scope
- Givens and assumptions
- Presentation by the consultants

Functional Analysis Phase:

- Reaffirmation of the functional objectives of the project

Creativity Phase:

- Idea generation
- Identification of additional options for harvesting of surface water
- Grouping ideas to clusters and re-define/create integrated solutions

Evaluation Phase:

- Discussion on the evaluation criteria
- Discussion on weightings of the chosen criteria

Action Plan Phase:

- Generation and agreement on follow-up actions

It is a fundamental success factor in group-based problem solving to have the appropriate participants who collectively have the appropriate knowledge and understanding of the issues and the possible options.

It is not possible to have everyone involved in the project to be present at the workshop. However, it is necessary to have key stakeholders represented, i.e. those who have strong influences on the direction of the project or those who would be most affected by the outcome of the project.

The participants of this workshop were carefully selected because of their influence or potential influence on the project. The selection was based on their ability to contribute positively to the issues to be discussed, and collectively share and own the outcome of the workshop. They would be responsible for following up the actions to be developed at the workshop. Importantly, the participants have to be at the appropriate decision-making level to contribute constructively in the workshop. In accordance with VM principles, major stakeholders of the project were invited and representatives from eight organisations from both the public and private sectors attended the workshop.

Outcomes of the VM Workshop

Information Phase

The purpose of this phase was to enable the entire team to have a thorough and precise understanding of the project. It covered key issues such as overview of the project, identification of client's requirements and areas of concerns, appreciation of project givens and constraints, challenge of project assumptions, and review of solutions proposed by the consultants. It is important for the workshop to define the scope of discussions through identification and clarification of the givens and assumptions of the project. Givens are fixed boundaries for the workshop to understand the constraints of the problem. Any discussion outside these boundaries would not contribute to the outcome of the workshop. Once agreed, givens are not to be challenged thereafter. Assumptions are estimated boundaries for the workshop. For certain issues of discussion, if it is not certain whether a particular concern is definite or not, the workshop will agree an assumption for the time being. Assumptions may be reviewed and challenged during and after the workshop. The following lists of givens and assumptions were identified, clarified, and agreed upon by the participants during the VM workshop.

PROJECT GIVENS

a.   It is necessary to explore the feasibility of maximising the collection and storage of surface water in City X.
b.   The focus of this workshop is on the collection of surface water in City X only, other methods of obtaining fresh water will not be considered during this workshop.
c.   ......

PROJECT ASSUMPTIONS

a.   This is one of the strategic studies in the territories; no timeframe is given for its implementation.

b.   The strategic storage of surface water collected is not a constraint and will be considered in the next workshop.

c.   ......

IDENTIFICATION OF KEY ISSUES

- *Existing surface water reservoirs and their gathering grounds*
- *Country parks*
- *Extensions to existing WGG*
- *River intakes and pumping station*

PROJECT CONSTRAINTS AND SPECIAL REQUIREMENTS

- *Extension of WGG*
- *Flood pumping areas*
- *Water quality*
- *Water treatment*
- *Environment, visual impact and traffic issues*
- *Health and safety*

Function Analysis Phase

The purpose of this phase was to identify the primary functions of the project, to clarify givens and assumptions of the project, and to explore opportunities for improving the project.

The primary functions of the project have been confirmed by the participants at the workshop as shown in Figure 10.3. The VM team focused on creation and evaluation of ideas to achieve these primary functions of the project throughout the workshop.

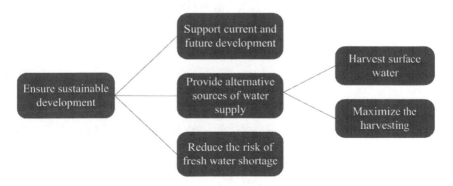

Figure 10.3 Function Hierarchy for Collection and Storage of Surface Water

Creativity Phase

The main task in this phase was to generate alternative solutions for accomplishing the primary function of the project. This phase answers the question: 'What else would perform the basic functions?' The VM team focused on creating ideas to enhance the project value through brainstorming. For better communication, the team were divided into four small groups comprising 4-5 participants. Through this exercise, the team generated a total of 35 ideas for achieving the required functions.

Having generated the alternative options, the team assessed the practicality of these options and put them into three categories: (P1) realistically possible, (P2) remotely possible, and (P3) fantasy. Table 10.2 shows the P1 ideas for maximising the collection of surface water.

Table 10.2 List of Realistic Ideas Generated at the Workshop

No.	Can We …… (Verb + Phrase)?	Cluster Covered
1	Collect rainfall at roof top	D
2	Extend and enlarge existing catch water	A
3	Collect higher proportion of flood water	C
4	……	

The realistically possible ideas were grouped into five clusters by the team, and a consolidated solution was proposed for each of the clusters. The heading of the clusters and the consolidated solutions are shown in Table 10.3.

Table 10.3 Five Clusters of Ideas and Integrated Solutions

A	Improve the harvesting of surface runoff from existing WGG, water collection and storage facilities
B	Extend WGG, water collection and storage facilities
C	Harvest flood water
D	Harvest storm water
E	Utilise raw water from existing irrigation and recreation reservoirs

Evaluation Phase

The main task in this phase was to set up criteria, and then evaluate and select the best solutions generated during the creativity phase by using the weighted evaluation technique. The first step of the evaluation was to establish a list of criteria, i.e. major factors to be taken into consideration during the evaluation of the alternative solutions. After extensive discussions among team members, the criteria for evaluation were agreed upon. Table 10.4 gives an overview of the criteria proposed and agreed upon by the VM team.

Table 10.4 Criteria Selected for Evaluation of Alternative Options

Category	Criteria
C1: Functionality	Amount of water that can be harvested
	Quality of harvested water
	Integration with water supply infrastructure
	Availability of energy
	Ease of construction
C2: Costs	Capital cost of project and associate transfer facilities
	Operation and maintenance (O&M) cost
	Opportunity cost of the land
C3: Environmental Impact	Potential noise impact
	Impact on water sensitive receivers
	Impact on drainage
	Potential visual impact
	Ecological impact
C4: Regulatory & Social Issues	Land status (for plants and associate facilitates)
	Compatibility with planning intentions/zoning
	Encroachment on country parks, archaeology & cultural heritage
	Miscellaneous regulatory issues
	Public safety
	Impact on land development
	Social implications
C5: Additional Benefits	Flood mitigation
	Slope stabilisation
	Other possible benefits

In order to assign weightings to these agreed criteria, to reflect their relevant importance in the evaluation process, a number of pair-wise comparisons were conducted by the team among the five categories. Based on the pair-wise comparisons, functionality and regulatory/social issues received the highest weighting of 10 out of 10. The weightings for cost and environmental impact were not too far away from the top two categories, whereas additional benefits were given the lowest weighting by the team.

In the process of assessing the relevant importance of the criteria, differences among the team members were observed. Discussions were hold to look into these differences and to exchange views; the value of the discussion was the insight into the issues, rather than the actual figures.

Having obtained the original weightings, the team reviewed them and finally agreed that it is more sensible to treat the first four categories of criteria with equal importance and therefore to assign equal weighting to them.

It is necessary to point out that this set of criteria and their weightings are the result of collective wisdom and extensive discussions among workshop team members. The process of assessment was democratic whereby different voices were clearly heard and issues discussed within the team, until a consensus was reached. It is crucial to understand that some of the criteria are more stringent than others. For example, some of the regulatory issues are compulsory and MUST be met by all options. The consultants should use the 'must be met criteria' to screen all options first and conduct a detailed evaluation among those that have met the minimum requirements of these criteria.

As the team felt that the five clusters of realistic ideas generated at the workshop were equally valuable and worth further studies, it was not necessary to evaluate and rank them. The team recommended that the consultants should take full consideration of the agreed criteria to evaluate alternative options during the subsequent stages of the study.

Based on the above criteria, the evaluation of alternative options would be undertaken using an evaluation matrix. This matrix would be prepared by the consultants when adequate information becomes available.

Development/Action Plan Phase

In order to ensure implementation of the feasible ideas generated during the workshop and criteria agreed by the stakeholders, a plan of necessary follow-up actions was developed during this phase. It is clear that the new options generated at the workshop represent some tangible benefits for the project as a whole. To ensure the development and implementation of these options, it was agreed by the team that the actions shown in Table 10.5 should be undertaken.

Table 10.5 Action Plan of Necessary Follow-up Actions

Action Required (Verb + Phase)		By	Date
A1	Collect and study information on the storm water drainage master plans and to explore the possibility of integrating with the surface water collection schemes.	XXX	XXX
A2	Consider possibility of integration with existing raw water transfer system for a) direct consumption by treatment works and b) storage in reservoirs.	XXX	XXX
A3	......		

Concluding Remarks

As a result of active participation and team efforts, the workshop achieved its objective. This conclusion was endorsed by the entire team at the end of the VM workshop. The VM workshop was attended by representatives of major stakeholders of the project. Critical issues and concerns were solicited and addressed to arrive at solutions acceptable to all stakeholders. It provided a comprehensive and holistic view of various aspects of the project and sought ideas and alternative solutions for value enhancement.

The team identified a large number of alternative solutions to maximise the harvesting of surface water in City X. The team also thoroughly discussed and agreed a list of criteria to be used in the evaluation of these options and the selection of the value-for-money options. The team of major stakeholders developed a common understanding about the project and consensus was reached among them in the selection and evaluation of suitable options for maximising the harvesting of surface water for the cost-effective provision of fresh water supply in City X.

The process and the outcome of the VM workshop were beneficial to all stakeholders of the project in developing a common understanding of the project objectives and concerns by various parties. It also helped ensure cost effective provision of freshwater supply at the subsequent stages by using a comprehensive list of agreed criteria in the evaluation of options.

It is of paramount importance for the consultants to build on the consensus developed among the stakeholders at the workshop, undertake follow-up actions raised by the team, investigate and develop additional schemes that can integrate the collection, storage and usage of the surface water in City X, and conduct a detailed evaluation of all options in respect of the agreed criteria in order to ensure that the project provides value for money.

Appendices

Appendix A – Rundown of one-and-a-half day VM Workshop

DAY ONE

8:30 am – 9.00 am	*Welcome Coffee/Tea Reception*
9:00 am – 10:30 am	*Workshop Process and Procedures*

- Introduction of participants
- Brief overview of the workshop process
- Confirmation of workshop objectives
- Ground rules of the workshop

Information Phase
- Overview of project objectives
- Latest design proposal
- Structural survey/proposed works
- Contractors' preliminary proposals
- Problems/opportunities
- Underlying stakeholder needs/expectations

10:30 am – 10:45 pm	*Break for Tea/Coffee*
10:45 am – 12:30 pm	*Analysis Phase*

- Key issues and concerns
- Givens/assumptions
- Function analysis
- Identification of value improvement opportunities
- Identification of base minimum/brief prioritisation

12:30 pm – 1:30 pm	*Lunch (and informal exchange of ideas)*
1:30 pm – 3:15 pm	*Creativity Phase*

- Identification of key areas to achieve objectives
- Generation of ideas to achieve objectives
- Generation of further ideas to achieve objectives

3:15 pm – 3:30 pm	*Break for Tea/Coffee*

3:30 pm – 5:00 pm	*Evaluation Phase*
	• Identification of evaluation criteria
	• Confirmation of option evaluation criteria
	• Evaluation and selection of ideas for further consideration
5:00 pm	*Close of Day One*

DAY TWO

2:00 pm – 3:15 pm	*Development Phase*
	• Development of options for evaluation against agreed criteria
	• Option evaluation and selection
	• Development of proposal communication document
	• Review of programme implications from selected proposals
3:15 pm – 3:30 pm	*Break for Tea/Coffee*
3:30 pm – 5:00 pm	*Reporting Phase*
	• Preparation of Action Plan
	• Resolution of outstanding issues
	• Review of workshop objectives
5:00 pm	*Feedback and Workshop Closes*

Appendix B – Rundown of one-day VM Workshop

Time	**Phase/Activity/Task**
8:45 am	*Welcome Coffee/Tea Reception*
9:00 am	*Welcome and Opening Remarks by Client*
9:05 am	**Information Phase**
	• Programme of the day
	• Participants' self-introductions
	• Organisation of the workshop
	• Introduction to the VM process
	• Agreement on workshop contract
	• Confirmation of VM workshop objectives
9:25 am	Presentation by BBV on 'Information Paper'
9:45 am	Comments/clarification on the information/ presentation

	• Givens and assumptions for the project • Opportunities for the project
10:30 am	*Coffee Break*
10:45 am	**Functional Analysis Phase** • Reaffirmation of the functional objectives of the project
11:00 am	**Creative Phase** • Introduction to idea generation • Identification/generation of new options
11:50 am	**Judgement Phase** • Classify ideas into P1, P2 and P3 categories • Group similar/relevant P1 ideas into clusters
12:30 pm	*Lunch (and informal exchange of ideas)*
1:30 pm	Review of ideas generated and collection of additional ideas • Make the clusters of ideas feasible solutions for implementation
2:00 pm	**Evaluation Phase** • Discussion and agreement on evaluation criteria and their weightings
3:00 pm	Evaluate and rank feasible options identified
3:30 pm	*Afternoon Tea*
3:45 pm	Evaluate and rank feasible options (continued)
4:15 pm	**Action Reporting Phase** • Identification and agreement on follow-up actions
5:15 pm	*Feedback and Workshop Closes*

Index